DESIGNING WITH OPERATIONAL AMPLIFIERS

THE  ELECTRONICS SERIES
BURR-BROWN

Wong and Ott · **FUNCTION CIRCUITS**

Graeme · **DESIGNING WITH OPERATIONAL AMPLIFIERS**

Graeme · **APPLICATIONS OF OPERATIONAL AMPLIFIERS**

Tobey, Graeme, and Huelsman · **OPERATIONAL AMPLIFIERS**

DESIGNING WITH OPERATIONAL AMPLIFIERS
Applications Alternatives

JERALD G. GRAEME

Manager, Monolithic Engineering
Burr-Brown Research Corporation

McGRAW-HILL BOOK COMPANY
New York   St. Louis   San Francisco   Auckland   Bogotá   Düsseldorf
Johannesburg   London   Madrid   Mexico   Montreal
New Delhi   Panama   Paris   São Paulo
Singapore   Sydney   Tokyo   Toronto

Library of Congress Cataloging in Publication Data

Graeme, Jerald G
  Designing with operational amplifiers.

  (The BB electronics series)
  Includes index.
  1. Operational amplifiers. 2. Electronic
circuits. I. Title. II. Series.
TK7871.58.06G73   621.3815′35   76-44537
ISBN 0-07-023891-X

Copyright © 1977 by Burr-Brown Research Corporation. All rights
reserved. Printed in the United States of America. No part of
this publication may be reproduced, stored in a retrieval system,
or transmitted in any form or by any means, electronic,
mechanical, photocopying, recording, or otherwise, without
the prior written permission of the publisher.

4567890   KPKP   7865432109

The information conveyed in this book has been carefully reviewed and is
believed to be accurate and reliable; however, no responsibility is assumed for the operability of any circuit diagram or inaccuracies in calculations or statements. Further, nothing herein conveys to the purchaser
a license under the patent rights of any individual or organization relating
to the subject matter described herein.

*The editors for this book were Tyler G. Hicks and Lester Strong,
the designer was Naomi Auerbach, and the production supervisor
was Frank P. Bellantoni. It was set in Caledonia
by The Kingsport Press. Illustrated by Lola E. Graeme.*

Printed and bound by The Kingsport Press.

CONTENTS

Preface ix

1. GENERAL AMPLIFIER TECHNIQUES 1

1.1 DC Error Reduction 1
    1.1.1 Continuous input offset voltage null 2
    1.1.2 Input bias current compensation 4
1.2 Boosting Input Impedance 7
1.3 Power Boosters 10
    1.3.1 Current boosters 10
    1.3.2 Voltage boosters 12
1.4 Boosting Full-Power Response 15
1.5 Continuously Variable Gain Control 19
1.6 Switched Gain Control 23
    1.6.1 Mechanical gain switching 24
    1.6.2 Electronic gain switching 26

2. AMPLIFIERS 31

2.1 Instrumentation Amplifiers 31
2.2 DC Motor Control Amplifiers 35
2.3 Clamping Amplifiers 39
    2.3.1 Zener diode–controlled clamping 39
    2.3.2 Voltage-controlled clamping 42

2.4 Controlled Current Sources 47
    2.4.1 Single-input current sources 47
    2.4.2 Differential input current sources 50

3. SIGNAL ANALYZERS 57

3.1 Comparators 57
    3.1.1 Specialized comparators 58
    3.1.2 Window comparators 64
    3.1.3 Reducing comparator hysteresis 69
3.2 Peak Detectors 76
    3.2.1 Basic peak detectors 76
    3.2.2 Specialized peak detectors 81
    3.2.3 Improving peak detector accuracy 84
    3.2.4 Improving peak detector speed 89
3.3 Voltage Discriminators 93

4. SIGNAL CONDITIONERS 99

4.1 Voltage Regulators 99
    4.1.1 General-purpose circuits 100
    4.1.2 Switching regulators 105
    4.1.3 Specialized voltage regulators 111
    4.1.4 Extending regulator utility 113
4.2 Active Filters 115
    4.2.1 Simplified state-variable configurations 116
    4.2.2 Digitally controlled active filters 118
4.3 Frequency Multipliers 121

5. ABSOLUTE-VALUE CIRCUITS 126

5.1 Single Amplifier Configurations 126
5.2 Precision Absolute-Value Circuits 130
5.3 Differential Input Absolute-Value Circuits 136
5.4 Absolute-Value Circuit Response Improvements 143
    5.4.1 Removing dc errors 144
    5.4.2 Extending absolute-value conversion bandwidth 145

6. SIGNAL GENERATORS 149

6.1 Wien-Bridge Oscillator 149
6.2 Square- and Triangle-Wave Generators 154
6.3 Ramp and Pulse Generators 161
6.4 Staircase Generators 164
6.5 Timing Circuits 169

7. COMPUTING CIRCUITS 174

7.1 Adders and Subtractors 174
7.2 Integrators 177
    7.2.1 Summing and noninverting integrators 177
    7.2.2 Extending integrator time constants 181

Contents vii

- 7.3 Differentiators 182
  - 7.3.1 Summing and noninverting differentiators 182
  - 7.3.2 Extending differentiator time constants 185
- 7.4 Multipliers and Dividers 187
- 7.5 Trigonometric Functions 193
- 7.6 Specialized Functions 196

8. DATA TRANSMISSION CIRCUITS 200

- 8.1 Two-Wire Transmitters 200
  - 8.1.1 General-purpose circuits 202
  - 8.1.2 Specialized two-wire transmitters 208
- 8.2 Voltage-to-Frequency Converters 212
  - 8.2.1 Moderate-precision configurations 214
  - 8.2.2 High-precision voltage-to-frequency converters 217

9. TEST AND MEASUREMENT CIRCUITS 224

- 9.1 Active Component Test Circuits 224
  - 9.1.1 Transistor test circuits 225
  - 9.1.2 Operational amplifier test circuits 227
- 9.2 Ohmmeters 231
  - 9.2.1 Conventional ohmmeters 232
  - 9.2.2 Ohmmeters for embedded resistors 235
- 9.3 Capacitance Measurement Circuits 237
- 9.4 Signal Measurement Circuits 241
  - 9.4.1 Digital voltmeters 241
  - 9.4.2 Ammeters 245
  - 9.4.3 Frequency measurement circuits 247
  - 9.4.4 Phase detectors 251
- 9.5 Electronic Thermometers 254

Glossary 258

Index 263

PREFACE

In the continuing evolution of analog electronics, circuit designers have originated and perfected a wide variety of operational amplifier realizations for electronic functions. These versatile amplifiers have become a basic component applied in varied engineering requirements for instrumentation, control, and simulation. Previous circuit applications have been described in the McGraw-Hill/Burr-Brown books *Operational Amplifiers: Design and Applications*[*] and *Applications of Operational Amplifiers: Third-Generation Techniques*.[†] Further developments in the operational amplifier art have generated a new set of implementations of electronic functions, and they are presented in this book.

Intended as a companion volume to the above publications, this book again presents operational amplifier applications in a form that permits ready adaptation to specific uses. Rather than limiting the applications to specific components and associated performance ranges, this book relays circuits in a general form with discussion of the circuit operation, critical component requirements, and related performance limitations. From such coverage, the reader can rapidly define the characteristics of components required to meet his specific application, or he can acquire insight into

---

[*] G. Tobey, J. Graeme, and L. Huelsman, *Operational Amplifiers: Design and Applications*, McGraw-Hill Book Company, New York, 1971.

[†] J. Graeme, *Applications of Operational Amplifiers: Third-Generation Techniques*, McGraw-Hill Book Company, New York, 1973.

circuit operation needed to modify the circuit for a related function. Use of the applications as presented does not require extensive electronics background because the expressions defining circuit operation and errors involve only simple engineering mathematics. From these expressions, one can select specific circuit components for the many uses throughout the engineering discipline without need for specialized electronic experience. To further aid general engineering use, a glossary is included at the end of the book.

So as to be a convenient reference in implementing electronic functions, the book is organized by circuit function. Applications for similar requirements are grouped together so that one can review alternative approaches in selecting the most appropriate technique for a given requirement. Grouped in Chapter 1 are general amplifier techniques for reducing error, boosting performance, and applying variable feedback networks. Chapter 2 summarizes the most recent implementations of specialized amplifiers for instrumentation, motor control, clamping, and current-source control. Signal analyzer groupings in Chapter 3 include comparators, peak detectors, and voltage discriminators. For signal conditioning requirements, Chapter 4 details selected voltage regulators, active filters, and frequency multipliers. Chapter 5 expands upon the widely used absolute-value circuits or precision rectifier approaches and means for their performance improvement. Included in Chapter 6 are newer techniques for generation of sine waves, square waves, triangle waves, ramp trains, pulse trains, staircase waveforms, and timed duration pulses. More powerful analog computation circuits for addition, subtraction, integration, differentiation, multiplication and division, and trigonometric and other functions are presented in Chapter 7. Two-wire transmitters and voltage-to-frequency converters for remote data acquisition and data conversion are covered in Chapter 8. Concluding the applications described are the test and measurement circuits of Chapter 9, which permit measurement of additional transistor and amplifier characteristics and measurement of resistance, capacitance, voltage, current, frequency, phase, and temperature.

I am once again grateful for the manuscript review by Donald R. McGraw, whose exceptional breadth of knowledge in the operational amplifier techniques resulted in many improvements. My appreciation is expressed to Burr-Brown Research Corporation, whose furthering of the analog-function technology provided the catalyst for derivation of much of the circuitry presented. My thanks to Fran Baker for her dedicated typing of the manuscript and occasional entertaining editing. Also, thanks to my wife Lola for again producing precise, visually pleasing illustrations and for providing the rewarding feeling of mutual involvement in preparing this book.

Jerald Graeme

DESIGNING WITH OPERATIONAL AMPLIFIERS

1
GENERAL AMPLIFIER TECHNIQUES

Of common use to the wide variety of operational amplifier applications are general techniques for enhancing performance and controlling closed-loop characteristics. Major circuit errors can be removed through techniques that compensate input offset voltage and input bias currents. Source loading errors can be avoided by making use of circuits that boost the input impedance presented by an operational amplifier circuit to a signal source. With a given operational amplifier an expanded range of functions is available through the use of output power boosting and through full-power response boosting. Utility is further extended by the ability to vary the effect of operational amplifier feedback networks. Generally, this involves continuous or switched variation of closed-loop gain, but the same control techniques also permit variation of response characteristics through switching a circuit function. Described in this chapter are techniques for achieving these results that can be applied to a wide range of operational amplifier applications.

1.1 DC Error Reduction

Input offset voltage and input bias currents create the major dc errors in operational amplifier circuits; so control of these characteristics can offer significant dc error reduction. For wideband operational amplifiers, the input offset voltage and its drift are generally quite large relative to more general-purpose amplifiers. The higher offset voltage is generally a result of optimization of response speed. Means are available for reducing that error voltage and drift to the same low levels available with low-drift operational amplifiers. Described below are techniques for using such low-drift operational amplifiers to provide a continuous offset null for wideband amplifiers or any other amplifier having high input offset voltage and drift. Following that description, techniques are presented for compensation of input bias current errors or compensation of the current itself. These latter techniques are derived for noninverting operational amplifier configurations where input currents produce more significant errors with the high impedance, or capacitive sources best monitored with these configurations.

1.1.1 Continuous input offset voltage null Input offset voltage and its drift are characteristically large for wideband operational amplifiers because of design compromises made for faster response. Similarly, low-drift operational amplifiers generally have slow response because their design has been optimized for low drift at the expense of response speed. This speed-accuracy compromise can be circumvented by combining a wideband amplifier with a low-drift one.[1] Together, the amplifiers form a low-drift wideband operational amplifier having the best characteristics of both, as well as greatly increased open-loop gain.

The two amplifiers are connected so that the wideband one controls response and the low-drift one determines input offset voltage. For inverting operational amplifier applications, the amplifiers are connected as in Fig. 1.1. As shown, the low-drift amplifier A_1 performs as an integrator controlling the voltage at the noninverting input of wideband amplifier A_2. The integrator senses the input offset voltage of A_2 and develops a correction voltage to remove it. Then, the only remaining input offset voltage will be that due to the low-drift amplifier A_1, which is

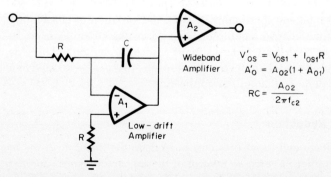

Fig. 1.1 Low input offset voltage and drift are provided by an integrator for wideband operational amplifiers to be used in an inverting configuration.

$$V'_{OS} = V_{OS1} + I_{OS1}R$$

and the composite amplifier drift is

$$\frac{\Delta V'_{OS}}{\Delta T} = \frac{\Delta V_{OS1}}{\Delta T} + \frac{\Delta I_{OS1}}{\Delta T}R$$

As connected, the continuous offset null circuit of Fig. 1.1 is a feedforward structure. High-frequency signals are fed forward around the bandwidth-limited amplifier to the wideband one; so the full bandwidth of A_2 is retained in the composite amplifier. The resulting gain and phase compensation requirements follow from feedforward amplifier theory.[2] At low frequencies, both amplifiers contribute gain as expressed by

$$A'_0 = A_{02}(1 + A_{01})$$

Accompanying the added gain is an additional phase shift with the potential for instability under feedback. To ensure frequency stability and optimize transient response, feedforward phase compensation is chosen for a continuous −6 dB per octave frequency response rolloff. This is achieved by setting the unity-gain crossover of the integrator at the same frequency as the first open-loop pole of the wideband amplifier. Because the integrator frequency response is controlled by its feedback elements R and C, these elements serve as the phase compensation for A_1. For unity integrator gain at the frequency of the first pole of A_2,

$$RC = \frac{A_{02}}{2\pi f_{c2}}$$

Here the first pole is represented in terms of the normally specified characteristics of open-loop gain A_{02} and unity-gain crossover frequency f_{c2}.

For noninverting or differential operational amplifier connections, an alternative approach is available for continuous input offset voltage null. The previous circuit is not suited for such connections because one amplifier input is committed to the offset correction signal. To leave both inputs available for signal or feedback, the offset correction amplifier can instead drive the offset null terminal of the wideband amplifier, as in Fig. 1.2. Again the low-drift amplifier A_1 senses the input offset voltage of the wideband amplifier and develops a nulling correction signal. Any change in V_{OS2} will be followed by an appropriate correction signal; so offset drift with temperature and time is continuously compensated for. The only remaining input offset voltage and drift will be due to the dc input errors of the low-drift amplifier:

$$V'_{OS} = V_{OS1} + I_{OS1}R_1$$

$$\frac{\Delta V'_{OS}}{\Delta T} = \frac{\Delta V_{OS1}}{\Delta T} + \frac{\Delta I_{OS1}}{\Delta T}R_1$$

4 Designing with Operational Amplifiers

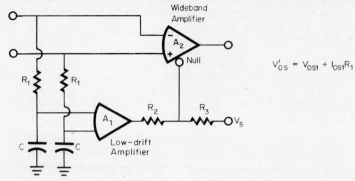

Fig. 1.2 For noninverting or differential input applications of wideband amplifiers, continuous input offset voltage null is provided by a low-drift amplifier that drives the wideband amplifier null control in response to any input offset.

Connections between the two amplifiers are made through input filters and a null control voltage divider. The low-pass filters provide phase compensation and block high-frequency signals that might unbalance the low-drift amplifier. Input interconnection polarity is determined by the signal phase relationship of the null terminal with respect to the inputs of A_2. If the inputs are correctly phased, offset will be driven to zero; otherwise it will be driven away from zero.

Also generally required for interconnection is the level-shifting voltage divider formed with R_2 and R_3. This divider ensures that the output voltage of A_1 will remain in its linear operating range even though the nominal potential of the null point is generally near one power-supply level. That supply voltage V_S is used to bias the divider for any required level shift.

Also provided by the circuit of Fig. 1.2 are increased open-loop gain and common-mode rejection. Low-frequency gain is boosted by the signal drive of the offset null point. As with the last circuit, the gain is added in a feedforward manner, and the added gain equals the open-loop gain of A_1 times the gain of A_2 from its null point to its output. Low-frequency common-mode error signals of the inputs of A_2 are also sensed by A_1; so a common-mode rejection signal is supplied to the null point as well. This feature is also highly beneficial because common-mode rejection is typically sacrificed with the other input characteristics in optimizing bandwidth.

1.1.2 Input bias current compensation The dc error created by operational amplifier input bias current is readily compensated in inverting amplifier applications. In these applications a compensation resistor is placed in series with the noninverting amplifier input, where it develops a compensation voltage with current flow in that input.[3] However, this technique offers little for most noninverting amplifier applications. A major advantage of noninverting configurations is the high input impedance they present to a source, making possible the monitoring of high impedance sources or holding capacitors. Flow of input bias current in such sources is often the dominant source of dc error for the noninverting operational amplifier. Described below are four techniques for compensating this error by impedance matching and by direct compensation of the input bias current.

By matching the resistance of the feedback network to the source resistance, compensating error voltages are developed by the two input bias currents, as illustrated in Fig. 1.3. The desired match is achieved by either inserting a compensation resistor R_3 or by choosing R_1 and R_2 so that their parallel combination equals the source resistance. Then, the dc error produced by the input current is associated with the difference between the two currents for typically an order of magnitude less error.

The need for the separate compensation resistor R_3 occurs when R_1 and R_2 cannot be made large enough for the desired compensation without sacrificing precision. Precise resistance is not as critical for the compensation of R_3 as it is for the gain set by R_1 and R_2. When R_3 is used, it should be capacitively bypassed as shown to preserve frequency stability. Without bypass R_3 forms a low-pass filter with the amplifier input capacitance resulting in increased phase shift in the feedback loop.

Where the source is capacitive rather than resistive, the same compensation cannot be used to remove the error of input bias current. Capacitive sources occur in many sample-hold circuits and peak detectors where a noninverting operational amplifier monitors the voltage stored on a capacitor. In such applications, the input bias current drains the capacitor, producing much of the circuit droop error. Here the need is for a direct compensation of the input bias current. This can be achieved with bootstrapping feedback or with an externally supplied compensating current.

For a voltage follower, the bootstrapping feedback is applied as in Fig.

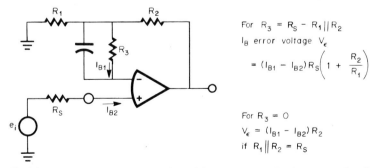

For $R_3 = R_S - R_1 \| R_2$
I_B error voltage V_ϵ
$$= (I_{B1} - I_{B2}) R_S \left(1 + \frac{R_2}{R_1}\right)$$

For $R_3 = 0$
$V_\epsilon = (I_{B1} - I_{B2}) R_2$
if $R_1 \| R_2 = R_S$

Fig. 1.3 For noninverting operational amplifier connections the dc error produced by input bias current can be greatly reduced by matching the feedback resistance to the source resistance.

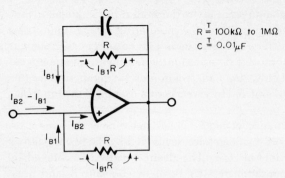

Fig. 1.4 The input current of a voltage follower can be reduced to the level of the input offset current through a bootstrapping feedback.

1.4. Here the circuit input current is reduced to essentially the difference between those of the operational amplifier at the expense of an output offset. Current from the inverting amplifier input flows through a resistor R to develop an output voltage $I_{B1}R$ for the bootstrap feedback. This voltage constitutes an output offset, but it biases the bootstrap feedback resistor to the noninverting input. As long as this voltage is much greater than the amplifier input offset voltage, it is essentially the voltage impressed on the bootstrap resistor. By making the bootstrap resistor equal to the feedback resistor, the bootstrapping current is made equal to I_{B1}. This current largely cancels the circuit input current because the two amplifier input currents tend to match.

To make use of this technique, it is desirable to null the amplifier input offset voltage, and it is necessary to bypass the negative feedback resistor. The nulling reduces error introduced by the amplifier input offset voltage in the feedback current, and it makes possible the use of a smaller bootstrap bias voltage for less output offset. Typically, an adequate bootstrap bias is developed with a resistance from 100 to 1,000 kΩ. Such resistance in the negative feedback loop must be heavily bypassed with about 0.01 μF to preserve frequency stability and transient response.

The input current compensation technique of Fig. 1.4 can also be applied to the general noninverting amplifier, although the associated output offset

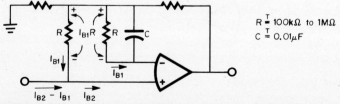

Fig. 1.5 A bootstrapping feedback can also be used to reduce the input current of a noninverting amplifier to the level of its input offset current.

can become large. Once again, the resistor R is inserted in series with the inverting input to develop a bootstrap bias voltage in Fig. 1.5. That voltage biases a matching resistor to supply the compensation current to the signal input. Compensation reduces the circuit input current to the level of the amplifier input offset current, as long as the amplifier input offset voltage does not significantly add to the bias on the bootstrap resistor. While this suggests that a large bootstrap bias voltage is desirable, the effect of that bias on the output offset must be considered. For the noninverting amplifier the bootstrap voltage is amplified by the circuit gain. Thus, it is desirable to carefully null the input offset voltage and keep the bootstrap bias small.

To avoid the offsetting effect of the bootstrap compensation technique, the input compensation current can be supplied by an external source.

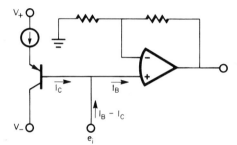

Fig. 1.6 Input bias current compensation from the base current of an external transistor ensures temperature tracking compensation.

For operational amplifiers using bipolar input transistors, a drift tracking compensation can be supplied from the base of an external transistor as in Fig. 1.6. By adjusting the current source, the compensation current I_C can be made to null the input bias current I_B. Thermal drift in the compensation transistor base current will tend to track that of the amplifier input transistor; so I_C will follow the drift of I_B.

Tracking is approximate because the compensation transistor cannot be matched to the amplifier input device, especially because they are of an opposite conductivity type. A pnp compensation transistor is required to compensate the base current of an npn input transistor, and vice versa. Additional compensation error arises from the voltage sensitivity of the current source. As signal swing varies the voltage on the current source, some change is induced in the current level. This represents a reduction in input impedance equal to a shunt of the current-source output resistance times the β of the compensation transistor.

1.2 Boosting Input Impedance

High input impedance contributes significantly to the accuracy with which operational amplifiers monitor signals, and numerous techniques are

8   Designing with Operational Amplifiers

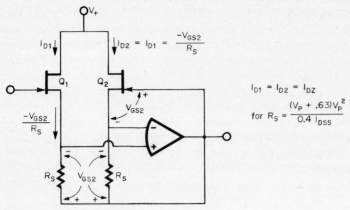

Fig. 1.7 Source-follower buffers at a voltage-follower input can be biased with resistors rather than current sources when a bootstrapping bias is used to remove signal swing from the resistors.

employed to boost that impedance.[3] Two others are presented here that provide bootstrapped FET (field-effect transistor) buffering of a voltage follower and permit connection of an inverting-only operational amplifier in noninverting configurations. The noninverting connection of such amplifiers achieves a dramatic increase in circuit input impedance because the signal then drives the amplifier input direct, rather than a summing resistor.

With FET buffering, the input impedance of a voltage follower is boosted to those levels limited by stray leakage resistance. A simple, high-performance approach to this buffer is the bootstrapped bias configuration shown in Fig. 1.7. In this circuit, the FET current bias is provided by resistors, but the FET currents are made immune to signal swing by bootstrapping bias from the amplifier output. Normally, source followers such as Q_1 and Q_2 would be biased from resistors connected to the negative supply. However, signal swing would then be directly impressed on these resistors, resulting in severe variation in the FET currents. Such current variations induce somewhat different changes in the gate-source voltages of Q_1 and Q_2, because the FETs cannot be perfectly matched. The result is a differential error signal that greatly degrades the gain accuracy of the voltage follower.

To avoid this error, source-follower buffers are often biased from current sources. This added circuit complexity is avoided with the bootstrap bias of Fig. 1.7 because the signal swing is removed from current-setting resistors R_S. Only the gate-source voltage of Q_2 is impressed on the source resistors, and this is sufficient for biasing. No other voltage bias is required, and it is this self-biasing capability of FETs that makes this technique possible. The bootstrapping connects the source resistor of Q_2 between the gate and source of that FET to fix its bias voltage. Essentially the same voltage is established on the other source resistor because near-zero voltage resides between the amplifier inputs. Thus the FET currents are

$$I_{D1} = I_{D2} = \frac{-V_{GS2}}{R_S}$$

No significant change in this current level occurs with signal swing because the voltage-follower output drives the source resistors with a signal that equals the one impressed at the input.

The exact level of I_{D1} and I_{D2} can be predicted from the FET pinchoff voltage V_P and zero bias current I_{DSS}. For a given value of R_S the biasing currents reach an equilibrium level where the gate-source voltage and source current of Q_2 are compatible. Generally, it is desirable for this current to be the zero temperature coefficient level I_{DZ}, where the gate-source voltage V_{GSZ} is independent of temperature. These zero-drift bias levels are related to FET characteristics by[2]

$$I_{DZ} \doteq \frac{0.4 I_{DSS}}{V_P^2}$$
$$V_{GSZ} \doteq V_P + 0.63 V \qquad V_P < 0$$

For the buffer of Fig. 1.7 a voltage equal to V_{GSZ} impressed on R_S will establish current equal to I_{DZ} if

$$R_S = \frac{(V_P + 0.63) V_P^2}{0.4 I_{DSS}} \qquad V_P < 0$$

Another technique provides high input impedance with single-ended input operational amplifiers such as most chopper-stabilized and feed-forward amplifiers, if the signal source can be floated. This approach is particularly useful for such inverting-only amplifiers because they cannot be connected in the noninverting configurations to achieve high circuit input impedance. Instead, these amplifiers can be connected in the non-inverting configuration of Fig. 1.8. In this circuit the signal is connected

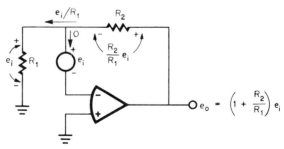

Fig. 1.8 High circuit input impedance with inverting-only operational amplifiers is achieved by connecting the signal source between the feedback network and the amplifier input.

directly to the high impedance input of the operational amplifier rather than to the summing resistor of an inverting configuration. This high impedance input draws very little current; so the signal source is lightly loaded.

Feedback forces the inverting amplifier input to essentially ground potential to produce the output signal shown. With zero voltage at that input, the feedback voltage on R_1 must equal the input signal e_i. Then, the feedback current must be e_i/R_1, and the signal on R_2 will be R_2e_i/R_1 for the output indicated. To preserve this response, care must be taken to avoid noise pickup on the feedback path through the signal source.

1.3 Power Boosters

Most operational amplifiers lack the output current and voltage capability needed to drive electromechanical devices, such as relays, motors, and speakers. For these applications a variety of circuits have been developed to boost output current and voltage.[3] Simplified current and voltage boosters are described below.

1.3.1 Current boosters
Complementary emitter followers provide the simplest bipolar current boosting for an operational amplifier. However, it is generally necessary to add considerable circuitry to bias such emitter-follower pairs in avoiding crossover distortion. That distortion occurs at the zero crossing of the output signal if one emitter follower must be turned off and the other turned on as the output current polarity reverses. Because some nonzero time is required to drive the transistors on and off, there is a portion of the output swing where neither transistor is on, and the output remains at zero.

Such distortion is removed by class A-B biasing of the complementary emitter followers as in Fig. 1.9. Diodes D_2 and D_3 provide a bias voltage

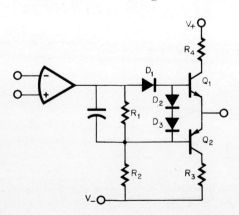

Fig. 1.9 Current sources are not required to ensure adequate base current supply under voltage swing when R_1 is added to provide a current path from the amplifier to Q_2.

which ensures that both transistors are on at the zero crossing. However, the resistor bias of such diodes is not generally adequate. If R_2 alone were used, it would be the sole source of base drive current for Q_2. The current available through R_2 varies with the output voltage, and at the negative swing extreme it is greatly reduced. Therefore, it is normally necessary to use a current source rather than a resistor to supply current to the biasing diodes.[3]

Instead, the resistor R_1 of Fig. 1.9 can be added so that the operational amplifier output current is also available to drive the base of Q_2. As the base current of Q_2 increases, it gradually diverts current from R_1 until the voltage on R_1 is zero. Any further base current demand will result in an amplifier output voltage below that of the base, and this reverses the direction of current in R_1. Now current from the amplifier is supplied through R_1 to the base of Q_2. Since the transition in R_1 current is gradual, no abrupt distortion is induced. The reversal in voltage on R_1 also reverse-biases D_1, which serves to protect Q_1 from damaging emitter-base breakdown. That diode may not be required if the reversed voltage on R_1 is sufficiently limited, as determined by the point of transition to reverse voltage.

Resistor R_2 is chosen to set the transition point, and R_1 sets the booster quiescent current. Generally, R_2 is selected for a transition point at an output voltage and current of about one-half the maximum desired. Beyond that point, little increase in voltage is required on R_1 to supply base current to Q_2 for the remaining desired output current. Greater voltage on R_1 would then only occur under output short circuit, where D_1 would be needed to protect Q_1. Also required under this short-circuit condition is an output current limit for each transistor; this is provided by R_3 and R_4. The latter resistors cause the transistors to saturate under excessive current flow. In this saturated mode the transistor current gains and the power dissipations are greatly reduced.

An even simpler low-distortion current booster is available if somewhat higher output resistance at low current is acceptable. This current booster has basically class B bias; so there is no quiescent drain from the booster. But the circuit is connected for continuous supply of output current through the zero crossing, as shown in Fig. 1.10. Near the output zero crossing, both transistors are off, but output current continues to be supplied through R_1 from the operational amplifier. No current boosting is then available, but it is not required for grounded loads when the voltage is small. As the output moves away from its zero crossing, load current is drawn through the resistor R_1, whose voltage gradually turns on one of the transistors to provide additional output current.

This transition will be gradual without crossover distortion as long as the amplifier is not forced into slew-rate limiting by the transition. Such a

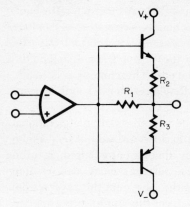

Fig. 1.10 Crossover distortion is removed from a class B current booster by a direct load current supply path through R_1.

rate-limited condition can result if the load resistance is small compared with R_1, such that the amplifier output swing would be much larger than that of the power booster during the transition. If the amplifier becomes rate-limited during the transition, it introduces a delay in booster output rise that appears as crossover distortion. To avoid this distortion, R_1 and the load resistance should be no smaller than about one-tenth the rated load of the operational amplifier. This permits current booster gains up to 10.

Because the operational amplifier output current is the source of bias to the emitter followers, the amplifier current limit serves to limit the booster current as well. If the operational amplifier output current reaches its limit level, the voltage on R_1 is also limited. That voltage determines the maximum bias for the transistors and their emitter resistors, and so the maximum booster current.

1.3.2 Voltage boosters To increase the output voltage swing of an operational amplifier, a gain stage can be added to the amplifier, or its power supplies can be bootstrapped from the output. Addition of a gain stage is fairly straightforward, and this stage can be greatly simplified by using the power-supply current drains of the amplifier for the bias and signal drive of this stage. By bootstrapping the amplifier power supplies, output voltage swing can be doubled[4] as in Fig. 1.11. In this circuit the voltages supplied to the amplifier power-supply terminals by the emitter followers are nominally one-half those of the actual power supplies. These amplifier supply voltages are bootstrapped from the amplifier output so that they track output swing. As a result, the amplifier output may swing to twice its normal rated voltage without the total voltage across the amplifier exceeding the maximum allowable. Essentially, this technique makes use of the amplifier total peak-to-peak output swing capability for each output swing polarity. In normal operation the maximum instantaneous output voltage of an operational amplifier is only one-half the total swing capability, because the other half of this capability is devoted to supporting a large biasing power-supply voltage. With the bootstrapping technique, no such bias is supported when it is not required for signal swing.

A potential failure of the above bootstrap operation is input overload. Fast-rising input signals could raise the amplifier inputs to high voltages before the amplifier output could respond to adjust the voltages at the amplifier supply terminals. Such overload would occur whenever the operational amplifier was forced into rate limiting unless input protection is incorporated. In the circuit of Fig. 1.11 resistors and diodes are added to clamp the amplifier inputs at the levels of the amplifier power-supply voltages.

Another way to increase amplifier output voltage swing is to add a gain stage. Normally, such a gain stage would be driven from the amplifier output, but a simpler circuit results if the added stage is driven as in Fig. 1.12. Here the signal, as well as the bias, is coupled to the gain stage through the power-supply current drains of the operational amplifier. No separate bias circuit is required. Bias for both the operational amplifier and the gain stage Q_3 and Q_4 is provided via Q_1 and Q_2. The latter transistors would normally be needed to supply reduced supply voltage to the operational amplifier. In this case, they also provide a source of biasing current and voltage level shifting to the bases of Q_3 and Q_4.

Signal current through the added stage is generated by operational amplifier output swing on resistor R_1. Any current supplied to R_1 must flow through one of the amplifier power-supply terminals, where it is

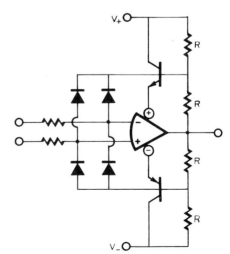

Fig. 1.11 Bootstrapping the power-supply voltages applied to an operational amplifier doubles output swing capability.

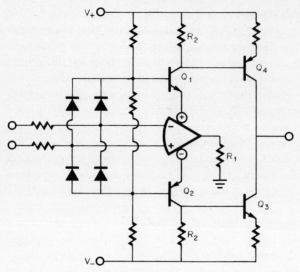

Fig. 1.12 An output voltage boosting stage without need for separate biasing is made possible by driving the stage with the power-supply line currents of the operational amplifier.

coupled to the gain stage. This coupling to the bases of Q_3 and Q_4 occurs with a voltage gain on the order of R_2/R_1. Voltage gain from there to the output is load-sensitive since the open-loop output resistance is quite high. Closed-loop output resistance, however, as well as any response nonlinearity, is greatly reduced by the feedback-loop gain.

Other characteristics of this circuit include the ability for current boosting, the lack of need for separate output current limiting, the potential for increased slewing rate, and a need for input protective clamping. By choice of resistors, the circuit can be made to boost output current as well as voltage. Output current limiting will often be provided by the internal limit of the amplifier, because its limited output current corresponds to a limited drive voltage to the gain stage. To preserve large-signal bandwidth with the higher voltage swings, a greater slewing rate is required. This too can be achieved with the circuit of Fig. 1.12. As described in the next section, the rate limiting of the operational amplifier can be avoided until higher frequencies by keeping its output voltage swing small. This is easily achieved by making R_1 small so that only small amplifier output voltage swing is required to develop its rated output current. Input protection is often required to protect the amplifier from the higher signal levels that may accompany higher voltage operation. If the input signals can be large enough to exceed the operational amplifier input ratings, the input clamp diodes and resistors shown should be added.

1.4 Boosting Full-Power Response

Operational amplifier large-signal bandwidth is specified as full-power response, and this bandwidth is lower than the small-signal bandwidth by typically a factor of 100. Specialized high-speed operational amplifiers greatly improve this ratio, but it can also be improved with general-purpose amplifiers, as described below. The techniques described boost full-power response by means of an external stage and by feedback modification that lowers the feedback factor without altering closed-loop gain.

The key to the first technique is a reduction in required amplifier output voltage swing. With the reduced voltage swing the slewing-rate limit of an amplifier is not encountered until a higher frequency. To restore full voltage swing to the load, an added gain stage is required, and normally this would require an accompanying increase in phase compensation. Such a phase compensation increase would then generally reduce full-power response to its previous level, or lower, but this can be avoided by decreasing the operational amplifier gain to prevent an overall gain increase.

This approach is incorporated in the circuit of Fig. 1.13. For simplified biasing of the added gain stage it is driven by the power-supply drain currents of the amplifier. In this way the quiescent currents of the amplifier provide quiescent bias, and the internal amplifier circuitry performs the voltage level shifting to the bases of the output gain transistors. Amplifier signal current is also conducted through one or the other of the

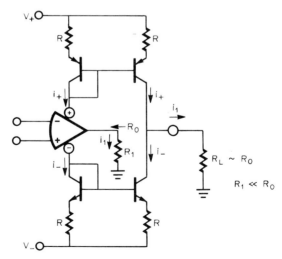

Fig. 1.13 Full-power response is boosted by lowering operational amplifier output voltage swing and restoring full swing to the new circuit output in a manner that does not increase the overall gain of the circuit.

power-supply terminals, where it drives the gain transistors. For the equal transistor emitter resistors shown, a unity current gain is established from the amplifier to the output transistors. Thus, a current i_1 supplied by the operational amplifier to R_1 will be matched by an equal current supplied to the load R_L. Then, the gain from the amplifier output to the new circuit output is R_L/R_1.

To maintain frequency stability without added phase compensation, the operational amplifier gain is reduced by a similar amount. This reduction is easily achieved by making R_1 small compared with the open-loop output resistance R_O of the amplifier. Then the amplifier open-loop gain is reduced by a factor of R_O/R_1, and the overall circuit gain is

$$A' \doteq A \frac{R_1}{R_O} \cdot \frac{R_L}{R_1} = A \frac{R_L}{R_O}$$

By making the load resistance comparable to the amplifier output resistance, circuit gain is not significantly changed, and additional phase compensation is not generally required. Full-power response is boosted because only a small amplifier voltage swing is required to supply rated current to the small resistor R_1. Typically R_1 can be selected so that only one-tenth the amplifier rated-output voltage swing is required to develop full voltage swing at the new circuit output. Where the various requirements above are not readily met, higher full-power response can still be attained if some additional phase compensation is placed in feedback around the operational amplifier.

For lower gain applications of externally phase-compensated operational amplifiers, full-power response can also be boosted at some sacrifice in noise performance and gain accuracy. This technique employs feedback connections that lower the feedback factor without increasing closed-loop gain.[5] Only a resistor and a capacitor need to be added to the inverting amplifier configuration for this full-power response boost illustrated in Fig. 1.14. Very little signal is impressed on the added elements R_3 and C; so they do not have a major influence on the amplifier output signal. Thus, the closed-loop gain remains essentially at the level set by R_1 and R_2.

However, R_3 and C do greatly alter the feedback ratio β of the circuit as represented by the $-1/\beta$ curve. The feedback ratio is that portion of the output signal fed back to the input, and it is calculated considering the impedance divider from output terminal to the amplifier input terminal.[2] At low frequencies where C appears as an open circuit, the feedback factor is unchanged from

$$\beta_L = \frac{R_1}{R_1 + R_2}$$

However, at the high-frequency extreme C appears as a short circuit and

$$\beta_H = \frac{R_1 \| R_3}{(R_1 \| R_3) + R_2} \doteq \frac{R_3}{R_2} \quad \text{for } R_3 \ll R_1 \text{ or } R_2$$

This reduction in feedback factor greatly eases the full-power response-limiting phase compensation, as can be seen from the response plot for $-1/\beta$. If β remained constant with frequency, it would intercept the magnitude response curve of the open-loop gain A at a point where the gain curve slope is quite steep. That steep slope correlates with high phase shift through the amplifier,[2] and such a phase shift is sufficient to cause oscillation. The condition for oscillation in a negative feedback system is a 180° phase shift where the feedback-loop gain is unity. That unity-gain point corresponds to the intercept of the $1/\beta$ curve and the gain magnitude curve, and this is the point at which phase shift must be limited.

Note that the decrease in β moves this critical intercept to a portion of the gain response curve that is much less steep, and thereby where the phase shift is much less. Frequency stability is then ensured without the need for heavier phase compensation, which would greatly decrease full-power response. Without the feedback factor decrease, phase compensation would have to be increased enough to maintain a -6 dB per octave gain slope past the intercept level of the R_2/R_1 line. The lighter phase compensation typically provides an order-of-magnitude increase in full-power response over that available with operational amplifier phase compensation selected for unity-gain applications. Small-signal bandwidth, however, is not increased, as indicated by the response of the closed-loop gain A_{CL}. This gain response declines with the open-loop gain response, following the intercept of the $-1/\beta$ curve.

Gain accuracy is commonly reduced by feedback factor decrease. However, this accuracy is largely restored by the greater open-loop bandwidth

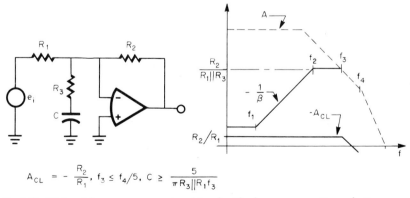

$A_{CL} = -\frac{R_2}{R_1}$, $f_3 \leq f_4/5$, $C \geq \frac{5}{\pi R_3 \| R_1 f_3}$

Fig. 1.14 Higher full-power response is achieved with phase compensation reductions permitted by a feedback factor decrease that does not alter closed-loop gain.

accompanying the decrease in feedback factor. Without the β decrease the bandwidth of the open-loop gain A would have to be greatly reduced by phase compensation. However, the high-frequency loop gain is preserved by the lighter phase compensation requirement. Low-frequency loop gain is maintained by the use of the coupling capacitor C, which prevents feedback factor reduction at low frequency. To optimize gain accuracy, the coupling capacitor should be carefully chosen to preserve frequency stability without unnecessarily degrading loop gain. The larger the capacitor, the less phase shift it introduces near the critical intercept point at f_3 but the more it reduces the lower frequency loop gain. Loop gain magnitude can be seen from the response plots as the difference between the A and the $1/\beta$ curves. Capacitor C combines with R_1 to create the $1/\beta$ response zero at f_1, and with R_3 to develop the pole at f_2. For frequency stability, f_2 should be a factor of 10 lower than f_3. Combining these various relationships defines the requirement for C as

$$C \geq \frac{5}{\pi R_3 \parallel R_1 f_3}$$

For good frequency stability also, f_3 should be about 5 times lower than f_4. Maximum benefit from this technique is achieved for R_3 much less than R_1 or R_2 so that the high-frequency feedback factor is significantly reduced. This lower value of R_3 would result in high gain to the amplifier input offset voltage, but that is prevented by the coupling capacitor.

Noise performance is more seriously affected by this full-power response-boosting technique. The input noise voltage of the operational amplifier is amplified by a gain of $-1/\beta$, which becomes large at high frequency in the curve of Fig. 1.14. As a result, high-frequency amplifier voltage noise is amplified much more than with the conventional operational amplifier feedback network.

An analogous full-power response-boosting modification can be made to the feedback network of a noninverting amplifier as in Fig. 1.15. This technique is shown for a voltage follower with components added to decrease the feedback factor at high frequency and permit phase compensation reduction. At low frequency the coupling capacitor C appears as an open circuit; so no feedback current is conducted through R_3, and the negative feedback factor is unity. At a higher frequency where C appears as a short circuit, the negative feedback factor is

$$\beta_{H-} = \frac{R_1 + R_2}{R_1 + R_2 + R_3}$$

In this case, feedback is also returned to the noninverting input, and the associated positive feedback factor is

$$\beta_{H+} = \frac{R_1}{R_1 + R_2 + R_3}$$

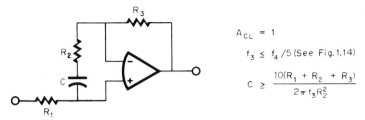

Fig. 1.15 For a voltage follower, increased full-power response is provided by phase compensation reductions made possible by a feedback factor reduction that does not alter closed-loop gain.

The net feedback factor at high frequency is the difference between the positive and negative levels

$$\beta_H = \frac{R_2}{R_1 + R_2 + R_3}$$

Thus, feedback factor at high frequency can be greatly reduced by making R_2 much less than R_1 or R_3. The resulting response curves are much like those of Fig. 1.14. Differences from those response curves include a unity feedback factor at low frequency and a unity closed-loop gain up to the rolloff at f_3. Closed-loop gain remains unity since very little signal is impressed on R_2 and C across the amplifier input. Very little signal current then is developed in the resistors R_1 or R_3; so the output signal essentially follows that at the input.

Feedback components required to provide the full-power response boost for the circuit of Fig. 1.15 are selected with the considerations outlined for the preceding circuit. Referring to the response plots accompanying that circuit, f_3 is again chosen to be about one-fifth f_4, and f_2 is set at about one-tenth f_3. This requires that

$$C \geq \frac{5}{\pi R_2 f_3}$$

While higher values of C help ensure frequency stability, they also reduce loop gain, and thereby gain accuracy, at intermediate frequencies. As before, the capacitor C also decouples high gain from the amplifier input offset voltage. High-frequency noise receives much greater amplification, as described with the last circuit.

1.5 Continuously Variable Gain Control

With potentiometers the closed-loop gain of operational amplifiers can be continuously varied over a wide range. In simplest form, the resulting gain varies nonlinearly with the potentiometer resistance. However, the gain can also be made to vary linearly so that turns-counting poten-

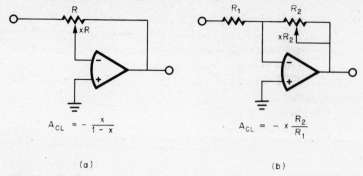

Fig. 1.16 The closed-loop gain of inverting operational amplifier configurations can be varied with only a potentiometer or, where linear control is desired, with a potentiometer and a resistor.

tiometer dials can be used for direct reading of gain setting. Described in this section are the potentiometer-controlled gain configurations commonly used with inverting, noninverting, and differential connections of operational amplifiers. Also presented is a circuit that provides bipolar adjustment of gain for variation of gain through both positive and negative levels.

For the inverting operational amplifier connection, the two basic potentiometer gain controls are illustrated in Fig. 1.16. As shown, gain can be controlled with the potentiometer alone or with a potentiometer and a resistor. When only the potentiometer is used, the control function is nonlinear as expressed in Fig. 1.16a. To achieve a linear control, the additional resistor of Fig. 1.16b is required. In both cases, the accuracy of the gain control function expressed is primarily determined by the accuracy of the potentiometer movement. Both circuits require careful interconnection of the potentiometer to the amplifier input to avoid noise pickup and stray capacitance. Stray capacitance introduces phase shift in the feedback loop and can cause oscillation.

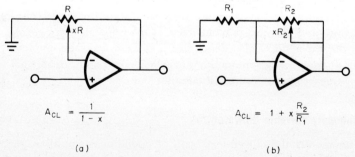

Fig. 1.17 Potentiometer control of noninverting amplifier gain is also possible if linear control is not required.

Noninverting operational amplifier configurations do not adapt as readily to linear gain control. The basic potentiometer-controlled gain connections are shown in Fig. 1.17. As indicated, both connections result in a nonlinear gain variation with potentiometer resistance. For linear gain control the noninverting amplifier can be followed by the variable-gain inverting amplifier of Fig. 1.16b, or a differential amplifier can be used. These differential amplifier circuits include that to be presented in Fig. 1.19 or specialized instrumentation amplifier circuits.[3] Generally, the differential amplifier circuits are preferable only where the phase inversion of the above inverting amplifier alternative is not acceptable. Otherwise, lower input impedance or greater circuit complexity results.

One of the means for attaining linear gain control with a noninverting amplifier response is to use the noninverting input of a difference amplifier having the desired gain control function. Two difference amplifier connections having potentiometer gain control are described below. The simpler circuit has a nonlinear gain control function, but the addition of a second operational amplifier provides linear control for the second circuit. In the simpler case, two resistors and a potentiometer are added to the basic difference amplifier connection[2] as in Fig. 1.18. Here the potentiometer performs as one segment of the two resistor tee networks of the difference amplifier. Because one potentiometer varies the effective resistances of both networks, common-mode rejection is not disturbed by the gain control. If two controls were used to separately vary the two net-

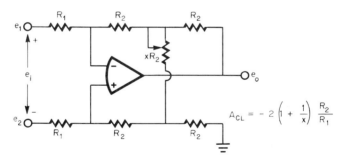

Fig. 1.18 A single potentiometer provides gain control for a difference amplifier without disturbing common-mode rejection when the potentiometer is connected as a common element of two difference tee networks.

works, high common-mode rejection would be difficult to maintain since it is very sensitive to the network resistance match. Common-mode rejection does still require close matching of like resistors, but tolerance error of the potentiometer only results in a gain error.

Linear gain control of a difference amplifier response can be developed by simply following a common difference amplifier with the gain-con-

22 Designing with Operational Amplifiers

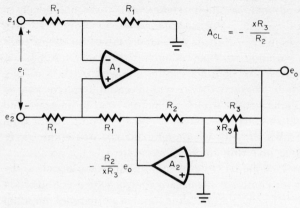

Fig. 1.19 Linear gain control for a difference amplifier results with a gain-controlled inverting amplifier inserted in the difference amplifier feedback loop.

trolled inverting amplifier of Fig. 1.16b, or to avoid cascading amplifiers a gain-controlled inverting amplifier can be connected in the feedback path of a difference amplifier as in Fig. 1.19. This inverting amplifier A_2 attenuates the output signal e_o to supply a feedback signal. Feedback forces the output signal of A_2 to equal the differential input signal, and it does so by developing the appropriate signal at the output of A_1. Relating the two amplifier output signals is the gain of the inverting amplifier, and the result is the gain control function expressed. Also added to the feedback are the offset voltage and phase shift of the inverting amplifier. The offset voltage is amplified along with that of A_1 by the net circuit gain. Phase shift introduced by A_2 can degrade frequency stability at lower gain levels.

Each preceding gain control technique permits variation of one gain polarity. Operational amplifiers, however, have bipolar voltage swing; so it is sometimes desirable to vary gain through both positive and negative

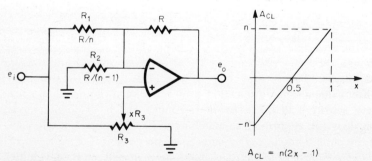

Fig. 1.20 Gain can be linearly varied through positive and negative levels with a single potentiometer that varies circuit operation from inverting amplifier to noninverting amplifier.

levels. A surprisingly simple operational amplifier connection results in such bipolar gain control by means of a single potentiometer[6] as illustrated in Fig. 1.20. In addition, the gain control function is linear. The potentiometer varies the signal applied to both the inverting and noninverting inputs of the amplifier to achieve both positive and negative gains. At one potentiometer extreme, where x equals zero, the noninverting input is connected to ground; so no signal reaches that input. This condition also holds the voltage across R_2 at zero so that it has no effect on circuit gain. Then, only R_1 and R conduct feedback current, and the circuit performs as a simple inverting amplifier with a gain of $-n$.

At the other extreme of the potentiometer range, where x equals 1, the input signal is connected direct to the noninverting input. Since feedback maintains near-zero voltage between the amplifier inputs, the inverting amplifier input will also reside at the potential of the input signal. Then, no signal is impressed on R_1, and feedback current flows only in R_2 and R. As a result, the circuit operates as a noninverting amplifier with a gain of $+n$. In between the two potentiometer extremes, feedback currents flow in both R_1 and R_2 for operation between that of an inverting amplifier and a noninverting amplifier. The polarity of the net circuit gain depends upon which operating mode is dominant, as determined by the potentiometer setting. Analysis reveals that the potentiometer setting linearly varies gain from $-n$ to $+n$ as plotted. The linearity of this gain control is solely determined by the potentiometer linearity, because resistance-ratio mismatch only affects gain magnitude.

Input resistance and offset voltage can, however, induce greater than normal error with this circuit. The circuit input resistance varies with the potentiometer setting, since the current drawn by R_1 varies with the setting of R_3. That input resistance varies from R_3, when R_1 conducts no current, to R_3 in parallel with R_1, where R_1 supplies its maximum feedback current. The resulting variation in loading on a source can produce an error variation in output similar to that of a nonlinear gain function. To minimize the input resistance variation, R_3 can be made much less than R_1, but this can make for a low value of input resistance. Larger than normal error also results from the gain provided to the amplifier input offset voltage. Both the inverting and noninverting amplifier paths amplify this offset even though they may not conduct signal current. As a result, the associated output offset is equal to two times the circuit gain times the input offset voltage.

1.6 Switched Gain Control

By switching the feedback networks connected to an operational amplifier, the closed-loop gain, or even circuit function, can be varied. Switching can be performed by either mechanical or electronic switches, depending upon the required control function and accuracy. With the mechanical types there is a choice of manual switches or automatic switching with relays. These switches have very low series resistance for minimal error, but they are relatively slow, and their wiring can degrade performance. Electronic switches are much faster and permit short wiring, but they have significantly greater series resistance. These varied switching limitations govern the circuit switching configuration. Generally, mechanical switches are connected in a configuration that minimizes the effects of noise pickup and stray capacitance from wiring. For electronic switches the circuit is configured to minimize error introduced by switch resistance.

1.6.1 Mechanical gain switching The wiring distance between mechanical switches and an amplifier that they control commonly results in noise pickup and stray capacitance. Because of associated stray signals and capacitance, mechanical switches should not be connected directly to an amplifier input. The high impedance of operational amplifier inputs makes these points susceptible to noise pickup, and the high signal sensitivity of the input results in an amplified noise error. In addition, the high impedance amplifier inputs do not shunt feedback resistance; so stray capacitance forms low-pass filters with the feedback resistance. From this, significant phase shift can be introduced into the feedback loop thereby degrading feedback stability. For these reasons, the preferred point for feedback switching is the amplifier output as in Fig. 1.21. The low output impedance of the amplifier greatly reduces the wiring sensitivity to noise pickup and the phase shift introduced by stray capacitance.

Noise immunity is still less than optimum with the connection of Fig. 1.21, since unused resistors will pick up noise and couple it to the input.

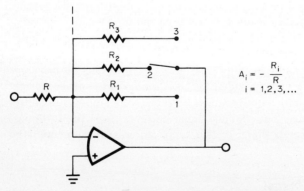

Fig. 1.21 The preferred configuration for operational amplifier gain control with mechanical switches incorporates feedback switching from the amplifier output where impedance is low, resulting in less sensitivity to noise pickup and stray capacitance.

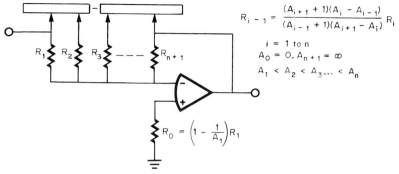

Fig. 1.22 With this switched-gain inverting amplifier connection there are no unused resistors that could pick up noise.

This effect is reduced by mounting the resistors close to the amplifier input so that the wiring to the amplifier is short. Further noise improvement results from capacitive bypass to ground of the switched end of each resistor, but this has other undesirable effects. First, it places increased capacitance load on the amplifier output with potential for frequency instability. Also, such bypass capacitors lower the feedback factor at high frequencies in the same manner described with the full-power response-boosting circuit of Fig. 1.14. As a result of the feedback factor reduction, closed-loop bandwidth and gain accuracy will be reduced unless phase compensation is switched with closed-loop gain.

Another alternative is available for avoiding noise coupling through unused resistors. With the alternative approach, all feedback resistors are used in each gain position so that there are no floating resistors that would pick up noise.[7] For an inverting amplifier this is accomplished with the connection shown in Fig. 1.22. Every resistor is connected in each gain position, and wiring capacitance loads the signal source and amplifier output. However, stray capacitance is not introduced at the amplifier input if the resistors are mounted close to that input. Gain is switched by means of a multiple-contact two-pole slide or wafer switch that connects the various resistors in parallel with either the input resistor or the feedback resistor.

While the resulting gain levels are determined by complex combinations of resistors, the selection of each resistance value is simple. For n different gain levels, n + 1 resistors are selected using the expression

$$R_{i+1} = \frac{(A_{i+1} + 1)(A_i - A_{i-1})}{(A_{i-1} + 1)(A_{i+1} - A_i)} R_i$$

where $i = 1$ to n, $A_0 = 0$, $A_{n+1} = \infty$, and $A_1 < A_2 < A_3 < \ldots < A_n$. First, a value is arbitrarily assigned R_1; then all other resistances are found from this starting point. For gains assigned in ascending order corresponding to

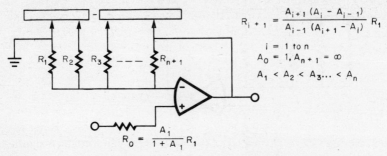

Fig. 1.23 Reduced noise sensitivity for a switched-gain noninverting amplifier is achieved by eliminating open switch positions to feedback resistors.

the number assigned each i of A_i, R_1 will be the smallest resistor, and positive values will result for all other R_i. Even the input current error compensation resistor R_0 can be selected for compensation at all gain levels with the expression

$$R_0 = \left(1 - \frac{1}{A_1}\right) R_1$$

An analogous configuration exists for noise reduction in gain-switched noninverting amplifiers, as shown in Fig. 1.23. In each switch position every resistor is connected either to ground or to the output; so no resistors are left floating to couple noise to the input. Wiring capacitance from the switch is shunted to ground or to the amplifier output. As before, resistor selection begins by choosing an arbitrary value for resistor R_1; then the remaining resistors are selected by the expression

$$R_{i+1} = \frac{A_{i+1}(A_i - A_{i-1})}{A_{i-1}(A_{i+1} - A_i)} R_1$$

where $i = 1$ to n, $A_0 = 1$, $A_{n+1} = \infty$, and $A_1 < A_2 < A_3 < \cdots < A_n$. Resistor R_0 is chosen for compensation of error due to input bias current by the expression

$$R_0 = \frac{A_1}{1 + A_1} R_1$$

1.6.2 Electronic gain switching While electronic switches lack the low series resistance available with mechanical switches, electronic switching of gain dramatically improves speed and reliability and is capable of direct interface with digital logic circuits. Electronic switches can also be mounted close to an amplifier so that wiring introduces less noise coupling and stray capacitance. This fact makes possible connection of electronic switches to the amplifier input for reduced error from switched resistance, as will be described.

Where the switch remains distant from the amplifier, electronic switches can be connected for gain control in the same configuration presented for mechanical switches in Fig. 1.21. The electronically switched equivalent of this circuit is Fig. 1.24, where MOSFETs (metal oxide semiconductor FETs) replace the mechanical switches. To ensure switch operation, the gate drive and substrate biases must be sufficiently greater than the maximum amplifier output swing. As before, different feedback resistors are connected to the amplifier output for different gain levels. However, in this case the ON resistance r_{ON} of the MOSFETs can introduce significant gain error, as seen by the gain equation

$$A_i = -\frac{R_i + r_{ON}}{R} \quad i = 1, 2, 3, \ldots$$

The initial error introduced by r_{ON} can be compensated for by adjustment of the R_i resistors, but r_{ON} varies with signal swing and temperature. Signal voltage at the output varies the gate-source voltages of the MOSFETs, resulting in a modulation of r_{ON} that introduces distortion.

Because electronic switches can be connected to an amplifier input without the long wiring of mechanical switches, it is possible to remove the distortion introduced by signal modulation of the switch ON resistance. This is made possible by the virtual ground of an inverting operational amplifier input and the switch connection of Fig. 1.25. Here the sources of the FETs are connected to that virtual ground; so there will be no signal swing across the gate-source terminals of the FETs. When an FET is in its ON state, the only signal swing across it will be the small voltage developed with the device ON resistance by the current through its input resistor R_i. That small voltage results in far less modulation of r_{ON} than experienced with the previous circuit. Further reduction in the

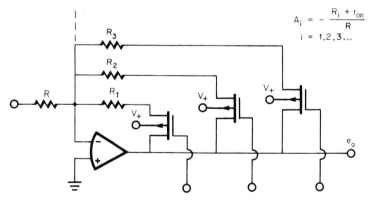

Fig. 1.24 Electronic gain switching can also be performed at the output of an operational amplifier, but a signal-modulated switch resistance will produce distortion.

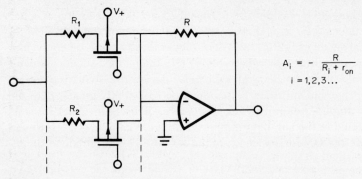

Fig. 1.25 By connecting gain-switching FETs at the virtual ground of an inverting operational amplifier, signal swing is removed from the FETs to avoid distortion from signal-modulated ON resistance.

ON resistance error is achieved by compensating adjustment of the R_i resistors, but the thermal variation in r_{ON} continues to develop error.

To totally remove error from switch ON resistance, the switches must be connected so that they conduct no signal current in the feedback path. This condition is very nearly true if the switches are connected in series with the amplifier input as in Fig. 1.26. With this arrangement, feedback networks are alternately switched to the amplifier input. Although this configuration is most sensitive to the noise pickup and stray capacitance of switch wiring, the critical wiring can be made short with electronic switches. Then, the only current conducted by the switches is the input current of the operational amplifier, which has an extremely small signal component. Thus, the error signal developed with the switch ON re-

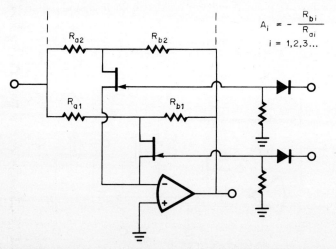

Fig. 1.26 In this gain-switching arrangement the error introduced by switch ON resistance is avoided by connecting the switches in series with the amplifier input rather than in series with the feedback network.

sistance is negligible, except perhaps at frequencies near the bandwidth limit of the amplifier. At such frequencies, open-loop gain is low so that the amplifier input capacitance has significant effect.

For this input switching arrangement, make-before-break switching is best. Otherwise, there will be a portion of the gain-switching cycle when no feedback network is connected to the amplifier. In this condition, the amplifier would swing toward one of its output saturation levels, resulting in a large output transient accompanying gain switching. For this reason, JFETs (junction FETs) are used for the make-before-break operation they provide. Alternatively, MOSFETs can be used if the switching is made sufficiently fast or if one feedback network is continuously connected to the amplifier. Fast switching limits the output transient if the switching time is too short for significant change in the amplifier output.

A continuously connected feedback network avoids the transient, except for that due to charge coupled through switch capacitance. Switched feedback networks are then connected in parallel with the permanently connected one, making resistor selection more complicated but offering a reduction in the number of switches. Even fewer switches can be used by making use of various combinations of switched feedback networks

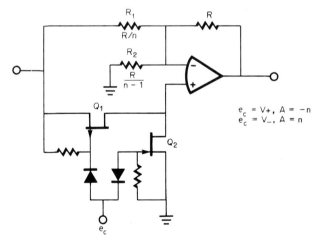

Fig. 1.27 Gain polarity can be switched without changing gain magnitude with a connection that switches from inverting amplifier to noninverting amplifier.

to set each gain level. By appropriate choice of resistors the gains can be binarily related to the switch drive signals, and operation will resemble that of a multiplying digital-to-analog converter.

Gain polarity can also be conveniently switched without changing gain magnitudes as in Fig. 1.27. This circuit is the switched-gain equivalent of the continuously variable gain configuration of Fig. 1.20. When the control voltage e_c is positive, Q_1 is off, and Q_2 is on to ground the noninverting amplifier input. In this state the voltage on R_2 is held at zero so that this resistor does not influence circuit gain. Only R_1 and R conduct feedback current; so the circuit performs as an inverting amplifier with a gain of $-n$. When e_c swings negative, switch states reverse. Then Q_2 is off, and Q_1 is on to connect the input signal to the amplifier noninverting input. Because the inverting input signal will follow that at its noninverting counterpart, the same signal appears at both ends of R_1. No current is then conducted by R_1, and feedback current flows only through R and R_2. The result is noninverting amplifier operation with a gain of $+n$. This gain will match the magnitude of the gain provided in the inverting amplifier state within the degree of match of the resistor ratio between R_1 and R_2.

For the switch configuration used in Fig. 1.27, switch error is minimized, and make-before-break switching is desirable. Steady-state switch error is negligible since the only current conducted through the switch ON resistance is the small input current of the operational amplifier. Switching transient error is avoided by using JFETs for their make-before-break switching operation, as described with the last circuit. If instead, break-before-make switching were used, both switches would be off for a portion of the switching cycle, leaving the amplifier noninverting input open-circuited. In that state the output would swing toward saturation, producing a transient during switching. With the make-before-break switching, both switches are on for part of the switching cycle. This does not short the input signal to ground, however, since the switch current is limited to the I_{DSS} levels of the FETs. To ensure that the appropriate FET remains off in a given state, the control voltage e_c must exceed the maximum signal swing by a voltage at least as great as the FET pinchoff voltage.

REFERENCES

1. W. Ott, Combined Op Amps Improve Overall Amplifier Response, *Electronics*, November 8, 1973.
2. G. Tobey, J. Graeme, and L. Huelsman, *Operational Amplifiers: Design and Applications*, McGraw-Hill Book Company, New York, 1971.
3. J. Graeme, *Applications of Operational Amplifiers: Third-Generation Techniques*, McGraw-Hill Book Company, New York, 1973.
4. S. Dogra, A Few Extra Components—741 Op Amps for High-Voltage-Swing Applications, *Electron. Des.*, April 26, 1974.
5. D. Kesner, Simple Technique Extends Op Amp Slew Rate, *EDN*, June 1, 1970.
6. J. Graeme, A Single Potentiometer Adjusts Op Amp's Gain over Bipolar Range, *Electron. Des.*, July 19, 1975.
7. E. Kacher and F. Fox, Eliminating Stray Signals in Remotely Gain-switched Op Amps, *Electronics*, May 2, 1974.

2
AMPLIFIERS

Operational amplifiers can be configured to form a variety of more specialized amplifiers including the instrumentation amplifiers, motor control amplifiers, clamping amplifiers, and controlled current sources of this chapter. In instrumentation amplifier configurations, they provide high-impedance differential inputs and precisely controlled voltage gain. As dc motor control amplifiers, operational amplifiers offer linear control of torque or speed. Clamping amplifiers permit linear amplification of signals up to voltage levels sharply controlled by zener diodes or control voltages. Current output connections of operational amplifiers result in amplifiers with transconductance gain or current gain rather than voltage gain.

2.1 Instrumentation Amplifiers

Voltage amplification requirements in instrumentation are often served by differential amplifiers commonly called *instrumentation amplifiers*. Basically, these amplifiers have a differential input and a feedback committed for voltage gain. While most operational amplifiers also have differential inputs, feedback is generally applied to one of these inputs, leaving only one signal input. When the feedback network is matched by an identical one to the second input of the operational amplifier, a differential signal input if formed, as in the elementary difference amplifier configuration.[1] However, this voltage feedback to the input results in a relatively low input resistance. The input resistance is only that of the two input resistors, which can introduce significant loading error. To retain high input resistance, the conventional voltage feedback can be replaced by current feedback as in Fig. 2.1. This circuit consists of an input emitter-follower pair biased from matched current sources and driving a feedback amplifier. The input is buffered from the shunting of the R_F feedback resistor by the β of the input transistors. This gives a differential input resistance of

$$R_I \doteq \beta(R_G \parallel 2R_F) \quad \text{if } R_G \gg r_e$$

where r_e is the dynamic emitter resistance. Common-mode input resistance is primarily determined by the two feedback resistors labeled R_F, and will be

$$R_{Icm} \doteq \frac{\beta R_F}{2}$$

Controlled gain is also provided by the current feedback as it supplies a current proportional to the input signal e_i. When an input signal is applied, it is essentially transferred to the gain-setting resistor R_G by the emitter-follower action of the input transistors. Initially this produces a current unbalance in the input transistors and, thereby, an unbalance in

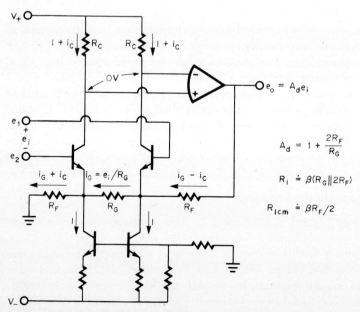

Fig. 2.1 Current feedback to a differential emitter-follower stage provides instrumentation amplifier operation.

the emitter-base voltages. The voltage unbalance results in an error in the voltage transferred to R_G. However, the current unbalance also creates a voltage between the operational amplifier inputs to generate a feedback correction signal. The operational amplifier returns its differential input voltage to zero by supplying a feedback current to one input transistor that equalizes the transistor currents. This feedback current flows in a resistor R_F, developing an output voltage of

$$e_o = \left(1 + \frac{2R_F}{R_G}\right) e_i$$

A matching resistor R_F is connected to the opposite emitter to balance the signal currents under common-mode swing and to eliminate the common-mode gain.

While this instrumentation amplifier configuration offers simplicity, it can have significant gain linearity and common-mode rejection errors. Gain nonlinearity is primarily a result of input transistor mismatch. Unless the input transistors track over their wide range of emitter current, unequal emitter-base voltages will be developed as signal and feedback vary the current level of these transistors. Then, a signal-dependent error is introduced that is not corrected by the feedback action described above.

Common-mode rejection is primarily limited by the matching accuracy of the R_F and R_C resistor pairs and of the input transistors. An input voltage common to the two input terminals is impressed on the R_F resistors; and if the resistors are unequal, they will initially conduct unequal currents. These currents flow through the input transistors to drive the operational amplifier inputs with any current difference. With such a difference, the amplifier will develop a feedback correction signal that represents an output error voltage. If the R_F resistors are perfectly matched but the R_C resistors differ, equal common-mode currents will still produce a signal at the amplifier inputs for an output error signal. Mismatch in the input transistor dynamic emitter resistances r_e will result in a differential error signal across R_G developed by the common-mode currents. Additional error is introduced by the output resistance mismatch of the input transistors and of the current-source transistors, but this is a second-order effect. Neglecting the latter, the common-mode rejection ratio (CMRR) is

$$\text{CMRR} = \frac{1}{\Delta r_e/R_F + \Delta R_C/R_C A_d + \Delta R_F/R_F A_d}$$

where A_d is the differential gain

$$A_d = 1 + \frac{2R_F}{R_G}$$

Much higher common-mode rejection can be achieved by configuring an instrumentation amplifier so that common-mode voltages are not impressed on the circuit feedback resistors. One such configuration, providing higher common-mode rejection at some increase in circuit complexity, is illustrated in Fig. 2.2. Once again, a differential emitter-follower pair, biased from matched current sources, constitutes the input stage. However, the feedback is returned to the input transistor collectors, where it is isolated from the common-mode input swing. Common-mode voltages are then impressed only on the collector-base junctions of the various transistors. Mismatches in the output resistances of like transistors will result in unequal currents under common-mode swing; however, these currents are very small since the transistors will have high output resistances. Thus, common-mode errors are greatly reduced from those of the previous circuit.

To achieve this, however, the gain linearity of the instrumentation amplifier is jeopardized by the flow of differential signal current in the input transistors. In the previous circuit, signal current was supplied to the gain-setting resistor R_G by the feedback path so that no current differential was developed in the input emitter followers. Current differential in these transistors develops a differential error voltage on their emitter-base junctions. That error voltage varies nonlinearly with the signal current, introducing distortion in the instrumentation amplifier output signal. To

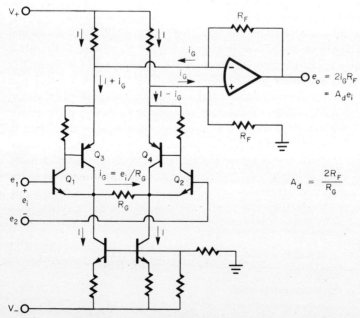

Fig. 2.2 Higher common-mode rejection is achieved with an instrumentation amplifier connection that avoids common-mode swing on the feedback resistors.

avoid this nonlinearity, the emitter followers Q_1 and Q_2 are bootstrapped by Q_3 and Q_4 in the complementary feedback pair connection. In this way, most of the signal current is diverted from the emitter followers to Q_3 and Q_4, avoiding the above signal-dependent error voltage.

The only signal current flowing in each emitter follower is the base current of its associated bootstrap transistor plus that generated in the emitter-follower bias. Signal variations in the current biases of Q_1 and Q_2 result from the signal-generated emitter-base voltage variations of Q_3 and Q_4 across the current-setting collector resistors of Q_1 and Q_2. However, the total signal currents in the emitter followers are greatly reduced, limiting nonlinearity to approximately 0.1 percent.

2.2 DC Motor Control Amplifiers

Often required in applications of dc motors is controlled torque or controlled speed as approximately provided in response to applied current or voltage. Torque motors develop a torque that is linearly related to the motor drive current, and in other applications controlled speed ideally results in response to the motor drive voltage. However, the motor armature resistance causes these torque or speed relationships to deviate from the ideal. Described in this section are motor control amplifiers that provide more nearly ideal response through a controlled current source, an armature resistance compensation, and a feedback speed control.

For controlled torque, it is desired to drive a torque motor with a current that is linearly related to a control voltage. Motor current, however, is not linearly related to an applied voltage since that voltage must support the motor emf. Operational amplifiers readily convert the control voltage to a current, with boosted output capabilities through use of a power booster, as in Fig. 2.3. Here the motor is connected inside the amplifier feedback loop and in series with the power-booster output. Thus, the current driving the motor will be linearly controlled by the input signal e_i. Linear control is assured by the closed-loop linearity of the amplifier. Under closed-loop conditions, feedback determines the voltage on resistor R_S, which senses motor current. To develop the required voltage on R_S, the amplifier forces the current i_m through the motor to R_S. That current, plus the current from R_2, produce a voltage on R_S that is linearly related to the input voltage e_i. Typically, the feedback current from R_2 will be small compared to i_m; so the voltage on R_S is essentially determined by i_m. The result is a motor current proportional to the input signal and an associated controlled torque of

$$\text{Torque} = ki_m \approx \frac{-R_2}{R_1 R_S} e_i \quad \text{for } R_S \ll R_2$$

where k is the torque constant of the motor.

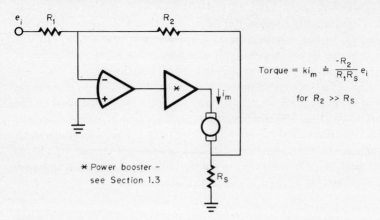

Fig. 2.3 For linear control of torque a torque motor is connected inside the feedback loop of a control amplifier so that motor current is linearly related to the control voltage.

To linearly control dc motor speed, it is necessary that the control voltage equal the back emf of the motor. However, the voltage applied to a motor supports both the back emf and the armature resistance drop. As a result, speed deviates from the desired linear relationship, especially under load, where armature currents are higher. Theoretically, the effect of armature resistance could be removed by addition of a negative resistance in series with the motor. That resistance would counteract the effect of the armature resistance.

In practice, this negative-resistance compensation can be achieved by means of positive feedback to the control amplifier[2] as in Fig. 2.4. With this circuit, the motor drive voltage is increased by a positive feedback signal from R_s to compensate for the armature resistance drop. Compensation is achieved by making

$$R_S = \frac{R_1 R_2}{R_3(R_1 + R_2)} R_A$$

Note that the required resistance of R_S is reduced for lower power dissipation by addition of R_2. This resistor increases circuit gain to the compensation signal without altering that supplied to the control signal e_i. Thus, independent control is available for the two gains. When R_S is set as expressed above, the voltage available to control the back emf of the motor, and therefore the speed, is directly proportional to the input control voltage. However the compensation feedback introduces the phase shift of the motor response into the amplifier feedback loop. Additional phase compensation is therefore required, and this is provided by capacitor C. That capacitor is selected large enough to stabilize the feedback loop,

Amplifiers   37

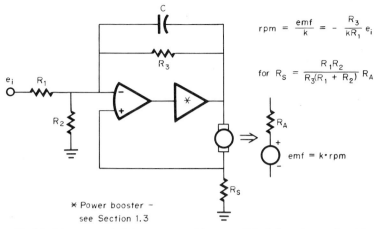

Fig. 2.4 Linear control of dc motor speed is accomplished through use of positive feedback to compensate for the signal voltage lost on armature resistance.

but with consideration for the accompanying reduction in control response speed.

Another means of dc motor speed control is tachometer feedback. By mounting a magnetic or optical sensor on a shaft-coupled wheel, a tachometer drive signal is provided. That signal indicates revolutions per unit time in terms of signal frequency, which can be transformed to a dc signal for comparison with a reference voltage. Deviations from the reference can be used to construct a correction signal for the motor drive.

For dc motors, this tachometer feedback speed control is implemented in the circuit of Fig. 2.5. Here the frequency of the tachometer signal is converted to a dc drive voltage by means of a one-shot and a differencing integrator. Each output pulse from the tachometer sensor triggers the one-shot formed with A_1 to produce a positive one-shot output state for a time t_1. Then the one-shot output returns to its negative state for some time t_2 until another trigger pulse is supplied by the sensor. A more detailed description of this one-shot operation is presented later with Fig. 6.15. The two one-shot output pulses have equal and opposite amplitudes as controlled by the clamping zener diode D_Z. As a result, the dc average of the one-shot output signal e_2 is proportional to the frequency of the motor revolution.

That average value is derived by the filtering of the integrator formed with A_2. Also, the integrator compares this average value against the reference voltage E_R set by the potentiometer R_6. If the average value of e_2 differs from E_R, there will be a net change in voltage on C_2 during a given cycle of e_2, and the motor drive voltage will be altered. At equilibrium, the charging and discharging of C_2 by $e_2 - E_R$ will be equal, and

38 Designing with Operational Amplifiers

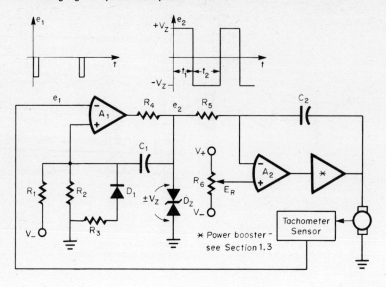

$$\text{rpm} = \frac{30}{t_1}\left(1 + \frac{E_R}{V_Z}\right), \qquad t_1 = (R_1 \| R_2)C_1 \ln\frac{2V_Z}{V_T}, \qquad V_T = \frac{R_2}{R_1 + R_2}V_-$$

Fig. 2.5 Tachometer feedback provides speed control for dc motors through frequency-to-voltage conversion and comparison of the resulting voltage against a reference.

no net change will result in the motor drive voltage. Then, motor speed will be

$$\text{rpm} = \frac{30}{t_1}\left(1 + \frac{E_R}{V_Z}\right) \qquad -V_Z < E_R < V_Z$$

In this equation V_Z is the zener voltage of D_Z which controls the amplitude of e_2. Since the average value of e_2 is constrained between V_Z and $-V_Z$, E_R must be similarly restricted to maintain speed control.

The charging and discharging of C_2 by e_2 does produce ripple on the motor drive voltage; so R_5 and C_2 must be made large to maintain constant speed. Additional reduction in the effect of ripple is provided by the momentum of the motor. That momentum also introduces phase shift in the tachometer feedback loop; so R_5 and C_2 must be selected with phase compensation considerations also. These elements should be large enough to assure stability, but they should not be excessive, such that they would unduly limit motor response to speed control changes.

To set the speed control range for the circuit of Fig. 2.5, the one-shot time period is fixed by choice of R_1, R_2, and C_1. These elements determine the charging time between trigger and reset. Triggering occurs when an input pulse swings far enough negative to drive the inverting input of A_1

below the threshold voltage established by R_1 and R_2 at the other input. That threshold voltage is

$$V_T = \frac{R_2}{R_1 + R_2} V_- > \frac{-V_{f1}}{2}$$

Note that V_T must be small enough that its presence at the noninverting input of A_1 does not forward-bias D_1. Any trigger pulse swinging below V_T will drive the output of A_1 positive, and this supplies positive feedback through C_1 to keep the output positive. A positive output condition will continue until C_1 has discharged through R_1 and R_2 to bring the noninverting input of A_1 back to the zero voltage level of the other input. Then the one-shot switches back to its negative state. The time t_1 spent in the positive state is

$$t_1 = (R_1 \parallel R_2)C_1 \ln \frac{2V_Z}{V_T}$$

In the negative state, C_1 recharges to its equilibrium level through D_1 and R_3, which shunt R_2 for shorter recovery time. During this recovery time, no trigger pulse can be received without resulting in timing error, and this limits the duty cycle range of the output pulses.

2.3 Clamping Amplifiers

Clamping amplifiers[3] or feedback limiters[1] provide amplitude limiting for signal clipping, signal squaring, and overload protection. One of the simplest clamping elements for these applications is the zener diode. The zener diode offers moderate precision where a fixed clamping level is desired. The feedback of amplifiers also makes possible voltage control of clamping levels. With voltage-controlled clamping, the clamp level is readily varied manually or electronically. Very low clamping voltages are then possible, and greater precision is often achieved. In this selection, various zener-controlled and voltage-controlled clamping amplifiers are described.

2.3.1 Zener diode–controlled clamping
Zener diodes connected across the feedback resistor of an operational amplifier, as in Fig. 2.6, will conduct when the amplifier output level reaches a certain voltage. This conduction shunts the feedback resistor and limits the output swing at the zener voltage of one diode plus the forward voltage of the other diode. Opposite polarity output swings are limited by the same diodes at an equal and opposite voltage level. By appropriate choice of the zener diodes, their zener voltage temperature coefficients can be made to partially

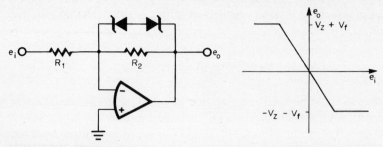

Fig. 2.6 In the simplest case zener diodes clamp amplifier output swing by shunting the amplifier feedback.

counteract those of their forward diode voltages for reduced thermal drift.

However, zener diodes used in this way impose serious limitations with their high capacitance, lack of a sharp turn-on characteristic, and high leakage current. Zener parasitic capacitance is comparatively high at typically 700 pF, and this capacitance can result in long clamp turn-on time and restricted signal bandwidth. For the zener to turn on, its capacitance must be charged through R_1, which is often a large resistance to preserve circuit input resistance. Signal bandwidth is limited by the capacitive shunting of R_2. When a zener diode turns on, it makes a transition from a high resistance to a low one; but because this transition is not abrupt, a sharp limiting is not achieved. Instead, the clamping is rounded. Even in its high resistance state the zener diode creates error with leakage current into the amplifier summing junction. When the zener diode is on, the clamp level it sets is still subject to thermal drift, since the zener will not likely be held at its zero temperature coefficient current.

All the limitations mentioned can be largely overcome with the biased zener clamp of Fig. 2.7.[4] The zener diode is continuously biased on, but it does not limit amplifier swing until the diode bridge connects it in the feedback path. Clamping occurs when the voltage on R_2 can support the zener voltage and forward-bias two bridge diodes. Positive output signals are clamped when D_3 and D_4 conduct to connect the zener diode across the feedback. Opposite polarity limiting is attained with the same zener when D_2 and D_5 conduct. Since the same zener limits both signal polarities, the output clamping will be symmetrical.

Continuous zener bias reduces clamp shunt capacitance, sharpens the clamping, and can often lower thermal drift. Lower drift is achieved by selecting the R_3 bias resistors for a zener thermal variation that is canceled by that of two bridge diodes. When the clamp is on, the zener current is approximately $(V_+ + V_f)/R_3$, or $(-V_- + V_f)/R_3$, where V_f is the forward voltage of the bridge diodes. This current calculation neglects the signal current from R_1, which is generally small compared with the required zener current.

Sharper clamping is achieved by avoiding the zener turn-on characteristic and leakage current. Clamp turn-on is now accomplished with the bridge diodes, rather than with zener diodes. Zener leakage current no longer affects signal current as the clamping level is approached. Leakage to the amplifier summing junction is replaced by the much smaller leakages of junction diodes D_2 and D_3.

Clamp capacitance is reduced by avoiding the charging and discharging of the zener capacitance. Only small voltage changes on the zener capacitance are now required, as produced by signal current flow in the already biased zener diode. Large voltage changes are isolated to the junction diodes, which have far lower capacitance than the zener diode. The equivalent clamp capacitance that must now be charged through R_1 is merely the combined capacitances of D_2 and D_3. Typically, this represents a 100:1 reduction from the capacitance of the basic zener feedback clamp; so turn-on time is dramatically reduced.

Even further reduced is the bandwidth-limiting capacitive shunt on R_2. Amplifier signals that do not turn on the clamp are not even shunted by the small bridge diode capacitance. When the bridge diodes are off, fixed voltages are established at one end of D_2 and D_3 by the zener and its bias resistors. Then, the only signal swing on the input shunting diodes D_2 and D_3 is the very small summing junction signal. This reduces the equi-

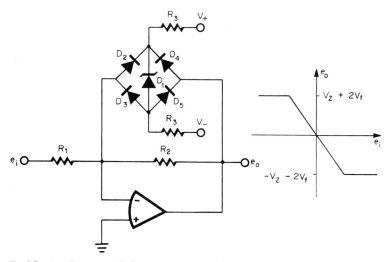

Fig. 2.7 Greatly improved clamping accuracy and speed are achieved with one zener diode that is continuously biased and that relies on signal diode switching to shunt the operational amplifier feedback.

valent capacitive shunt of R_2 to $2C_f/A$, where C_f is the forward capacitance of a junction diode and A is the open-loop gain of the amplifier. This capacitance is negligible in comparison to other parasitic capacitances.

2.3.2 Voltage-controlled clamping With operational amplifiers, signal clamping can be controlled by a voltage rather than a zener diode. This makes possible precise clamping at low voltage levels not available with zener diodes and provides means for amplitude modulation of square waves. Sharper clamping is also provided with the voltage-controlled techniques as signal current is isolated from the clamped reference voltage. However, clamping is generally limited to signal bandwidths significantly lower than the actual amplifier bandwidth. Described below are clamping amplifiers that permit clamping at one or more levels which can be fixed, manually varied, or electronically controlled.

Single-level clamping is provided by the simple circuits of Fig. 2.8. These circuits are similar to the elementary diode clamp except that the diode is enclosed in an operational amplifier feedback loop to remove the signal loss otherwise required to forward-bias the diode. By having the high-gain amplifier supply the diode voltage, clamping is extended to very low levels. Clamping of the more positive portion of a signal is provided by the circuit of Fig. 2.8a. When the input signal e_i is less than clamp level e_L, the amplifier output is driven positive to reverse-bias the

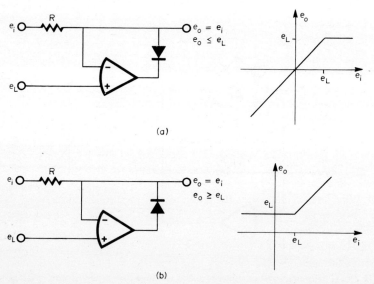

Fig. 2.8 Sharp clamping at a voltage-controlled level is attained by use of an operational amplifier to supply the clamp diode bias voltage and to buffer the clamp level control source.

feedback diode. Then the signal is passed through the resistor to the output without limiting. In this state the resistor can represent a relatively high output resistance so that output loading can induce error. Further error will be created by any current drawn through the resistor by the amplifier input; so it is desirable to use an operational amplifier having high input resistance under input overload.

When e_i reaches the level of the clamp voltage, the amplifier output swings negative, forward-biasing the feedback diode and clamping the output. In this state the amplifier acts as a voltage follower to the control voltage e_L, while signal current from e_i is absorbed by the amplifier output. The result is the input-output response shown. For clamping of the lower portion of a signal, the diode is simply reversed as in Fig. 2.8b.

Sharp clamping is achieved with these simple circuits by isolating the clamp control voltage from signal currents. No clamp level variation is created by input signal increases beyond the clamp level because the amplifier buffers the control signal e_L from signal currents. The control signal drives the high impedance source. Alternatively, the amplifier buffering can be assigned to the input signal by reversing the input connections of e_i and e_L along with a reversal of the feedback diode. The identical input-output response results, but e_i is buffered rather than e_L. However, this does result in signal current flow to the control voltage source; so it must then be low impedance to retain sharp clamping.

Clamping performance with the circuits of Fig. 2.8 is limited by the amplifier dc input errors and by overshoot during switching transitions. The circuit switches to the clamping state when the current reaching the feedback diode reverses and not necessarily when the input signal reaches the level of e_L. Actual switching is offset by $V_{OS} + I_B R$. To remove this switching offset, the amplifier null control can be adjusted, but this results in an output offset voltage. Both the switching offset and the output offset can be removed by combined use of the amplifier null control and a dc correction signal summed into the amplifier input.

Output overshoot is developed on these clamp amplifiers because of the time required for the operational amplifier to switch the feedback diode state. Until this switching is completed, increases in e_i beyond the clamp level will continue to reach the output. The amount of overshoot is determined by the rate of change of the input signal and the amplifier output voltage. Slewing-rate limit determines the time required for the operational amplifier output to swing from its saturation level occupied in the nonclamped state to that voltage required to forward-bias the diode. Techniques for reducing this overshoot-producing switching time are the same as those described in Chapter 5 for improving the response of absolute-value circuits.

To limit both the positive and negative swings of any signal, the two

44 Designing with Operational Amplifiers

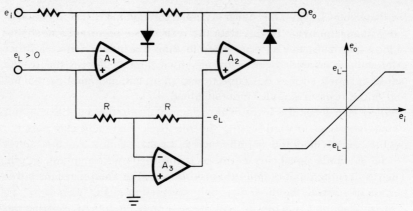

Fig. 2.9 Symmetrical positive and negative limiting is achieved by combining the two circuits of Fig. 2.8 with an inverter.

circuits of Fig. 2.8 can be cascaded with two different control voltages applied. However, in many limiter applications it is desirable to use a single control voltage to set both a positive and a negative clamping level. This can be achieved by addition of an inverter to the cascaded circuit as in Fig. 2.9. Positive limiting is provided by A_1 when e_i reaches e_L, and negative limiting is produced by A_2 when e_i reaches $-e_L$. Retained by this circuit are the overshoot and high output resistance of the separate limiter circuits described above.

Alternative circuit approaches provide the positive and negative clamping of the previous circuit with greatly reduced output resistance and overshoot at the cost of somewhat increased circuit complexity. One such circuit makes use of half-wave precision rectifiers[5] as in Fig. 2.10. Normally, the precision half-wave rectifier passes only one polarity of signal; so it essentially limits the signal at zero. This limit level is shifted away from zero by summing a limit control signal e_L to the two rectifier circuits of Fig. 2.10. Opposite polarity limits are produced by a single control signal because of the inverter action of A_1. In addition to limiting e_i, A_1 inverts the signal passed to the second limiter formed with A_2. This inverted signal is limited on the opposite swing by the same control voltage e_L. Also developed by e_L are signals at the outputs of A_1 and A_2. To remove such signals, a summing amplifier A_3 is added, and the resistors are ratioed as shown. With the output summing amplifier, low output resistance is assured.

Output overshoot during switching will still occur with this circuit, but it is greatly reduced by the feedback clamping diodes of the precision rectifier. These diodes directly shunt the feedback paths of A_1 and A_2 to limit their output swings when the amplifiers are disconnected from their resistive feedback paths. Without these added diodes, the outputs of A_1 and A_2 would swing to their saturation levels, as occurred with the previous circuits in this section. Output saturation introduces a switching delay that is avoided by the clamping diodes. In addition, the clamping diodes greatly reduce the signal swing during switching transitions; so switching time and thereby overshoot are reduced. Other errors of this circuit result from resistor mismatches and amplifier dc input errors. Means for removing the effect of dc errors with the precision rectifiers are described in Chapter 5.

To avoid the resistance mismatch error of the above circuit, an alternative approach is available that reduces the number of resistors to be matched to two, if reduced output current is acceptable. This circuit uses three separate amplifiers to alternately control the three separate segments of a symmetrical limiter input-output response as in Fig. 2.11.[6] Controlling the linear segment is voltage follower A_1, which has a series output resistor that makes it possible for the other two amplifiers to override the control of A_1. Either of the other amplifiers can draw current from the output of A_1, developing a voltage on R_1 that saturates this amplifier. In the saturated mode, A_1 can no longer control the circuit output voltage.

When e_o reaches a voltage equal to e_L, it drives the output of A_2 negative, forward-biasing D_1 to connect this amplifier to the circuit output. In this state, A_2 acts as a voltage follower that attempts to hold the output voltage

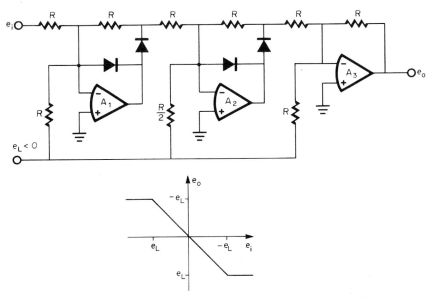

Fig. 2.10 Reduced overshoot and output impedance result with a symmetrical limiter formed by two half-wave precision rectifiers and a summing amplifier.

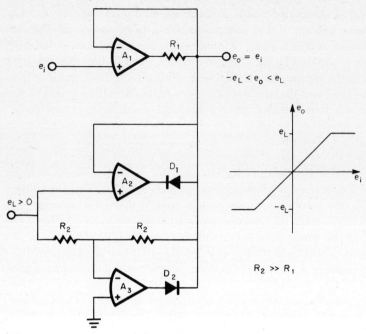

Fig. 2.11 Symmetrical clamping is produced by three amplifiers which alternately control the circuit output for the three segments of the input-output response.

at e_L. To do so, it must override the follower action of A_1, which it does by supplying current to R_1 to saturate A_1. When e_i drops below e_L, the output of A_2 swings positive, reverse-biasing D_1. This disconnects A_2 from the circuit output so that A_1 again has output control.

Similar limiting action is provided on negative output swings by the half-wave precision rectifier formed with A_3. If the output signal reaches a level of $-e_L$, the feedback resistors on A_3 couple a signal to the input of that amplifier that forces its output to swing positive, forward-biasing D_2. Then A_3 performs as an inverter to hold the circuit output at $-e_L$ by overriding A_1.

Again, output control is diverted from A_1 by conducting current through R_1. That resistor must be large enough to permit this control but not so large that it excessively limits the current A_1 can supply to the circuit output. For uniform output current capability in the three circuit modes, R_1 should be set to saturate the output of A_1 when one-half the rated amplifier output current is conducted through that resistor. Then A_2 and A_3 will only have to devote one-half their output current capabilities to saturate A_1, and the remaining half is available for the circuit load. Also, R_2 must be made large compared with R_1 so that it does not consume too much of the remaining output current capability of A_1.

2.4 Controlled Current Sources

Most electronic instrumentation is composed of circuits which produce a signal voltage from a control voltage. However, signal currents derived from control voltages can sometimes provide more straightforward solutions for instrumentation requirements. Such voltage-controlled current sources are useful in testing and for driving certain loads. In transistor testing, controlled current sources provide simple programmable current biasing.[3] Resistance measurement is simplified with a current rather than a voltage test signal, as contact resistance will not affect the signal supplied from a current source. Current output is also needed for meter drive, dc torque motor drive, and process control instrumentation.

In current output configurations, operational amplifiers perform as signal-controlled current sources for a wide range of applications. These varying applications may require unipolar or bipolar output currents, single-ended or differential inputs, grounded or floating loads or sources, and varying degrees of accuracy. Many of these requirements are met by one or another of previously described current-source configurations.[3] Additional flexibility is provided by the configurations described in this section, particularly by the differential-input-controlled current sources.

2.4.1 Single-input current sources Where a single input rather than differential inputs is required, precise control of output current is simply achieved. Especially simple configurations result if the input signal source can be floated or if it is a current. Illustrated below are two circuits for developing a bipolar output current for grounded loads from floating voltage sources. Also described is a current inverter.

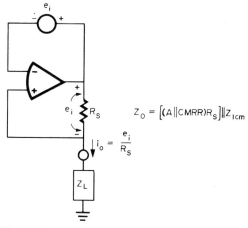

Fig. 2.12 Bipolar output current is precisely controlled by a floating signal source with a circuit that is floating on the load signal voltage.

Signal sources that can be floated are readily isolated from signal voltages developed on the current-source load, ensuring high output resistance. The signal source is bootstrapped to isolate it from load voltage swings as in Fig. 2.12. Here the entire current-source circuit floats on the load signal. Any signal on Z_L is sensed by the noninverting input of the operational amplifier, causing the amplifier output to drive the signal source and current-setting resistor R_S to follow load voltage variations.

In doing so, the amplifier develops small input error voltages associated with common-mode rejection and gain error. These error voltages and the common-mode input impedance of the operational amplifier produce output current error related by an output impedance of

$$Z_O = [(A \parallel CMRR) R_S] \parallel Z_{Icm}$$

Also introducing small errors are the input offset voltage and bias current of the amplifier for an output error current of $I_B + V_{OS}/R_S$. Otherwise, output current is controlled by e_i and R_S alone as indicated. Feedback forces the voltage on R_S to equal e_i in order to maintain zero voltage between the amplifier inputs. Thus, the circuit output current equals

$$i_o = \frac{e_i}{R_S}$$

As expressed in the last equation, the transconductance gain of this circuit is controlled solely by R_S. While this is convenient, it does impose a limit to the level of transconductance which can be precisely set. Higher levels of transconductance require small resistance values, which are not available with precision resistors. For higher transconductance, voltage gain can be added to increase the voltage set on R_S. This is readily done

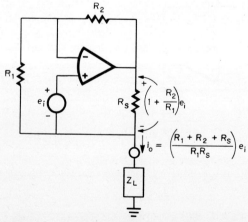

Fig. 2.13 A controlled current source with higher transconductance results with a noninverting amplifier bootstrapped from the circuit load.

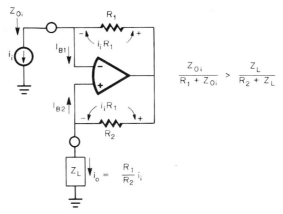

Fig. 2.14 Combined positive and negative feedback on an operational amplifier result in an inverting current amplifier.

where signal sources can be floated using the circuit of Fig. 2.13. Essentially, this circuit is a common noninverting amplifier configuration bootstrapped from the load voltage. As a result of the amplifier gain, the voltage developed on R_S is amplified for increased output current without lowering R_S. Additional output current flows to the load from the resistor R_1 for a net output current of

$$i_o = \frac{R_1 + R_2 + R_S}{R_1 R_S} e_i$$

It is often desirable to invert the polarity of a signal current or to precisely amplify it. This can be accomplished using common operational amplifier techniques by converting the current to a voltage, inverting or amplifying that voltage, and then reconverting it to a current. However, what is really required is an inverting current amplifier such as that of Fig. 2.14. Input current i_i flows in the feedback resistor R_1, where it is converted to a signal voltage. That voltage is impressed on R_2 since there will be essentially zero voltage between the amplifier inputs. Resistor R_2 reconverts the signal to current i_o for supplying the load. The result is a load current opposite in polarity from that of the input current source and amplified by the ratio of the two circuit resistors.

Small errors are created in the load current by the amplifier dc input errors and by amplifier signal characteristics. From the amplifier input offset voltage and input bias currents, an output error current is created equaling

$$I_{B2} - I_{B1} \frac{R_1}{R_2} + \frac{V_{os}}{R_2}$$

Signal-related errors determine the output impedance of the current source from the signal source output impedance Z_{Oi} and the amplifier common-mode input impedance Z_{Icm}, gain error, and common-mode rejection error. The result is

$$Z_O = Z_{Oi} \parallel Z_{Icm} \parallel [(A \parallel CMRR)R_2]$$

Care must be taken to ensure the frequency stability of the circuit of Fig. 2.14 since it employs positive feedback. For stability, the net feedback factor must be negative, as achieved by making the feedback factor to the inverting input greater than that to the noninverting input:

$$\frac{Z_{Oi}}{R_1 + Z_{Oi}} > \frac{Z_L}{R_2 + Z_L}$$

This ensures that the net impedance in the input circuit is positive. By itself, the current source has a negative input impedance of

$$Z_I = -\frac{R_1}{R_2} Z_L$$

However, the net impedance in the input circuit remains positive as long as Z_{Oi} is greater than the magnitude of Z_I above.

2.4.2 Differential input current sources The inherent differential input capability of operational amplifiers can be extended to controlled current sources for rejection of common-mode signals. Simpler differential input configurations have lower input resistance, but high-impedance differential inputs can be formed using differential amplifier structures. The simplest form requires only one operational amplifier but has restricted load voltage swing and requires precise resistance adjustment to achieve high output impedance. That circuit is the same as a common single-input controlled current source[1] except that both amplifier inputs are signal-driven as in Fig. 2.15.

Linearly controlled load current is developed through positive feedback with this circuit. Input signal e_2 supplies current directly to the load, but this current is altered by the load voltage e_L. Additional current is supplied to the load through the R_2 positive feedback path. Controlling this current is the voltage developed by e_1 and e_L at the amplifier output. Fortunately the effect of e_L on this current opposes that on the current supplied by e_2, and the two effects can be made to cancel. Cancellation occurs when the positive and negative feedback networks are matched as shown, and the current supplied to the load is then essentially independent of the load voltage.

Low-frequency performance of this current source is limited by resistance matching and the dc input errors of the operational amplifier.

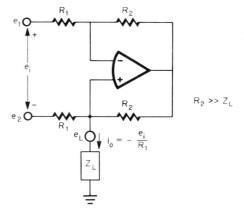

Fig. 2.15 A differential-input-controlled current source is formed with matching feedback networks to the two inputs of an operational amplifier.

Extremely careful matching of the R_1 and R_2 resistor sets is necessary to maintain high current-source output resistance and high common-mode rejection. For perfect resistor matching, the output resistance approaches R_1 times the common-mode rejection ratio of the amplifier, and the common-mode rejection ratio of the current source equals that of the amplifier. The input offset voltage and input offset current of the operational amplifier V_{OS1} and I_{OS1} produce an output current error of I_{OS1} plus V_{OS1}/R_1. Generally, the total error current can be compensated for by adjusting the amplifier null, unless R_1 is small.

For the circuit of Fig. 2.15, load voltage swing is greatly limited by the requirement that the current source be loaded by an impedance much less than that of R_2. With Z_L much less than R_2, the positive feedback will be significantly less than the negative feedback as needed to keep the amplifier in a linear mode. However, with Z_L much less than R_2, the amplifier output swing is greatly attenuated before reaching the load. This restricts the load voltage to a fraction of the amplifier output voltage swing capability.

Significantly improved load voltage swing capability can be attained at the expense of an additional operational amplifier as in Fig. 2.16. This circuit consists of a typical difference amplifier connection bootstrapped from the load voltage. To visualize circuit operation, consider the input and output of follower A_2 to be grounded. In this condition the circuit would be a difference amplifier with load resistor R_S. Well-controlled voltage would be established on this resistor by the closed-loop characteristics of the difference amplifier, such that the current in that resistor would be $-e_i/R_S$.

To make use of this controlled current, the entire difference amplifier is bootstrapped on the load voltage. Voltage follower A_2 is added to isolate the load from undesired current. While A_2 does couple positive feedback

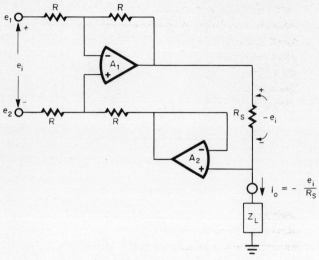

Fig. 2.16 A difference amplifier becomes a differential input current source when bootstrapped from the load voltage.

signals to A_1, the associated positive feedback factor will always be less than that of the negative feedback. Positive feedback is reduced by the dividing action of R_S and Z_L; so the voltage swing limitation of the preceding circuit is largely avoided. However, the precise resistor matching requirement is not removed with the circuit of Fig. 2.16. The equal-valued resistors must be extremely closely matched to avoid degradation of common-mode rejection and circuit output impedance.

A somewhat simpler configuration for a differential-input-controlled current source is that of Fig. 2.17. As shown, the circuit consists of opposing FET current sources that are controlled by high-gain feedback around an operational amplifier.[7] The difference in FET currents produces the output current, and this difference current is controlled by summing feedback to the input from the current-sensing source resistors R_S. At feedback equilibrium, the sum of the two feedback signals is directly related to the differential input signal and

$$i_o = -\frac{ne_i}{R_S}$$

Differential inputs and power-supply rejection are provided by an attenuator network at the inverting amplifier input that matches the feedback network connected to the other amplifier input. This is analogous to having the matched input and feedback networks connected to an operational amplifier to form the common difference amplifier.

To simplify biasing and improve large-signal bandwidth, the FET gates are driven from the amplifier supply terminals rather than from its output terminal. Quiescent biasing for the FETs is provided by the amplifier quiescent current drains, and no level-shifting bias is required from the amplifier to the FETs. Large-signal bandwidth is improved by the reduced output swing required from the amplifier in this configuration as explained in Sec. 1.4. Only a 1 V amplifier output swing on a 100 Ω R_L is required to draw typical rated output current. This current is drawn through the supply terminals for maximum drive to the FETs. The lower amplifier output swing is not as greatly bandwidth-limited by the amplifier slewing-rate limit. Optimum bandwidth is achieved by choosing R_L small enough to limit swing but not so low as to excessively lower the amplifier gain. Large-signal bandwidth is then commonly limited by the amplifier common-mode swing rate limiting.

The result of this circuit structure is a unique combination of capabilities for a simple controlled current source. A bipolar output current is provided that can drive grounded or floated loads. This current is precisely controlled by the input voltage within the accuracies of the resistors shown and within the gain-bandwidth limitations of the operational amplifier. Accuracy limitations of the FETs are overcome by the feedback, except

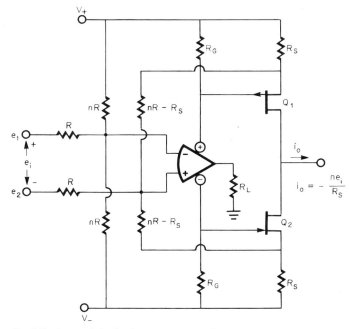

Fig. 2.17 Summed feedback to an operational amplifier from the sources of complementary FETs results in a bipolar output current controlled by a differential input voltage.

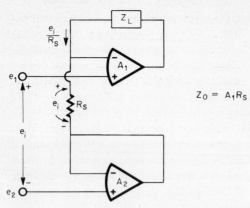

Fig. 2.18 Controlled current-source operation having high output impedance, high input impedance, and high common-mode rejection is achieved by connecting the circuit load in the feedback loop of this differential amplifier.

for the small gate-drain leakage currents. Output current is limited to the I_{DSS} levels of the FETs, but this can be boosted using the transconductance multiplying technique sometimes applied to common FET-controlled current sources.[3] Output impedance is feedback-multiplied from that of the FETs to the practical limit imposed by stray and parasitic effects of approximately 10^{12} Ω and 10 pF.

By virtue of the differential inputs, common-mode signals are rejected by a CMR (common-mode rejection) adjustable to over 90 dB, but very precise resistor matching is again required. Primary CMR limitations are the accuracies of the resistor ratios and matches shown, except for the noncritical match between the resistors labeled R_G. CMR adjustment is made by trimming the input resistors. Prior to this adjustment any desired nulling of dc offset should be performed by trimming the resistors denoted nR. The differential inputs also accommodate floating sources.

Each of the preceding differential input current sources has relatively low input impedance set largely by the circuit input resistors. Where higher input impedance is required, other differential amplifier structures can be adapted to current-source operation. One such adaptation, suitable where the circuit load can be floated, is the circuit of Fig. 2.18. Here the two inputs present the high common-mode input impedances of operational amplifiers to a signal source. Also provided by this current source are very high output impedance and common-mode rejection.

Load current is precisely set by feedback without need for the close resistance matching of the preceding current sources. Only R_S determines the circuit transconductance, within the gain and common-mode errors of the operational amplifier. Controlling the voltage on R_S is amplifier feedback, which maintains near-zero voltage between inputs of each amplifier. As a result, the voltage on R_S is forced to equal the differential input signal e_i. To establish this voltage, feedback supplies a current of e_i/R_S through the load.

Added to this load current are several error currents. Contributing dc errors are the input bias current of A_1 plus a current equal to the difference between the two amplifier input offset voltages divided by R_S. Signal error in the load current results from the load voltage swing divided by the gain of A_1. This error defines a current-source output impedance equal to A_1R_S. Additional signal error current is produced by the input common-mode voltage swing through the error voltages it creates at the inputs of the two amplifiers. Generally, the common-mode error voltages of the two amplifiers will partly cancel, resulting in a common-mode rejection for the current source that is greater than that of the individual amplifiers.

Where loads must be grounded, the circuit of Fig. 2.18 can be modified. The output of A_1 can be grounded if the current source is operated from a floating power supply. A simpler alternative is available for a differential input current source with high input impedance using an instrumentation amplifier[1,3] in a current output configuration, as in Fig. 2.19.

Forming this current source is a differential input amplifier stage consisting of A_1 and A_2 followed by the current source presented earlier in Fig. 2.15. The lower output resistance and load voltage limitation of that previous circuit also affect this configuration, but common-mode rejection is improved. Reduced sensitivity to common-mode errors results from a

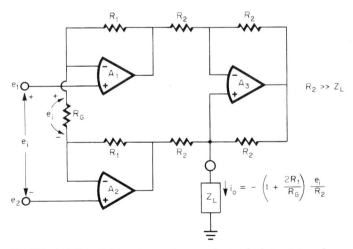

Fig. 2.19 A differential input current source having high input impedance and capable of driving grounded loads is provided by an instrumentation amplifier connected in a current output configuration.

differential gain provided by A_1 and A_2. These amplifiers transmit input common-mode voltage to the output current source with unity gain, but the differential signal can be greatly amplified by those amplifiers, so that it is less sensitive to the common-mode errors of the output circuitry. Relatively little common-mode error is introduced by A_1 and A_2 in comparison to that resulting from the mismatch between the R_2 resistors. Essentially no common-mode error is produced by the mismatch in the R_1 resistors.

REFERENCES

1. G. Tobey, J. Graeme, and L. Huelsman, *Operational Amplifiers: Design and Applications*, McGraw-Hill Book Company, New York, 1971.
2. L. Van Allen, DC Motor Control Circuit Cancels Armature Resistance, *Electronics*, April 12, 1973.
3. J. Graeme, *Applications of Operational Amplifiers: Third-Generation Techniques*, McGraw-Hill Book Company, New York, 1973.
4. J. Graeme, Continuing Biasing Improves Clamping Amplifier, *Electronics*, July 11, 1974.
5. J. Smith, *Modern Operational Circuit Design*, John Wiley & Sons, Inc., New York, 1971.
6. L. Drake, Long Time Constant Oscillator Uses Precision Clamps, *EDN*, December 20, 1974.
7. J. Graeme, Controlled Current Source Is Versatile and Precise, *Electronics*, May 16, 1974.

3
SIGNAL ANALYZERS

Numerous characteristics of signals can be monitored with operational amplifier circuits. Among the more common characteristics of interest are voltage and current amplitude, frequency, and phase, as measured with the circuits in Sec. 9.4. Additional information is provided by comparators, peak detectors, and voltage discriminators as realized with operational amplifiers in this chapter. Comparators provide indications of when signal amplitude exceeds a reference level or when the amplitude is within a certain range. Such information is used in limit testing, alarm indicators, and feedback timing control. Uses for peak detectors arise where signal information is not measurable in terms of root mean square (rms) or average value. Such is true for irregularly shaped or nonrepetitive signals. Voltage discriminators perform signal selection on the basis of amplitude and pass the signal selected to the circuit output.

3.1 Comparators

One of the most common applications of operational amplifiers is voltage comparison. While design-committed comparators are available, their variety is much more limited than that of operational amplifiers. As comparators, operational amplifiers provide a choice of low drift, high speed, low input current, low cost, etc. Some of the functions that can be performed with operational amplifiers as comparators include the specialized comparators and window comparators of this section. Also described are means for reducing parasitic comparator hysteresis error, which is often the dominant source of comparator error.

3.1.1 Specialized comparators

Among the many functions performed by comparators are signal squaring, pulse detection, hysteresis control, and alarm indication. Specialized comparator configurations for these purposes are illustrated below. Signal squaring is one of the simplest roles for a comparator because it only requires high gain to transform a rounded signal into sharp pulses. However, at very low signal amplitudes the signal detection is degraded by the input offset voltage of the comparator and any dc level accompanying the signal. If the combined dc error voltage is greater than the signal swing, the signal will not trip the comparator.

However, by means of dc feedback the comparator threshold can be automatically adjusted to remove the effects of offsetting dc voltages and permit detection of very small ac signals. This is accomplished with a basic operational amplifier open-loop gain circuit[1] as in Fig. 3.1. By appropriate choice of R and C, the gain at the signal frequency will not be affected by their feedback, while the gain at dc is removed. To block the dc gain, the capacitor charges to an appropriate voltage that shifts the comparator threshold to the level of the ac signal. If initially the threshold is not reached by the ac signal, the comparator output remains in one state charging C through R until the threshold reaches the ac portion of the signal. This threshold adjustment continues until the average value of the comparator output signal equals the dc portion of the input signal E_I plus the input offset voltage of the amplifier. The resulting average value of the output signal is approximately zero for small E_I; so the output signal

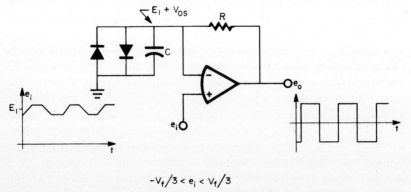

Fig. 3.1 Squaring of very small ac signals, normally prevented by the presence of comparator input offset voltage or other dc error voltages, can be restored by dc feedback.

will be a nearly symmetrical square wave. With this technique, rounded ac signals with amplitudes of only a fraction of a millivolt can be converted into sharp square waves with 10 V amplitudes.

However, the circuit of Fig. 3.1 is quite slow in adjusting its threshold because of the gain-preserving requirement for a large time constant associated with R and C. These elements must be made large enough that they do not feedback-limit the comparator gain at the signal frequency. Also, feedback stability requires a large time constant. Feedback phase shift is introduced by these elements, as in the basic differentiator configuration,[1] and it may be necessary to add resistance in series with C to maintain frequency stability. With the required large time constant, changes in capacitor voltage take significant time especially during circuit turn-on. At turn-on, a large overload voltage can be developed on C, and clamping diodes are added to limit this effect. This addition also restricts input signal range to a fraction of a forward diode voltage V_f.

Another means of automatic threshold adjustment can be used to reject noise or dc level shift in pulse detection. As often occurs with pulse-coded information, the amplitude of pulses is not well controlled, and comparator detection of such signals must be set for a minimum amplitude. This condition is particularly true in data transmission, where data pulses can be significantly attenuated by transmission lines or by radio transmission losses. Also accompanying such transmission is noise, and comparator detection must be adjusted for a threshold above the maximum anticipated noise level yet low enough to detect the minimum signal amplitude. Because of this compromise, noise margin is typically set at its minimum level and held there even though signals may be well above the minimum level anticipated.

Far better noise margin can be achieved by means of a comparator that automatically adjusts its threshold to be compatible with signal amplitude. Specifically, the comparator threshold can be adjusted on the basis of preceding pulse amplitudes to hold the comparator threshold above the noise level yet within the range of following pulses. This is accomplished by using a peak detector to set the comparator threshold[2] as in Fig. 3.2.

In this circuit, A_1 operates as a peak detector, and A_2 performs the comparator function to square and remove noise from data pulses. The comparator threshold e_T is set by the peak detector output, which stores the maximum previous level established by a signal pulse. Ideally, the threshold is set at the midpoint of the pulse transition. To avoid the noise on both the high and low signal states, the resistors R_1 and R_2 are chosen to decay the voltage on capacitor C to one-half its peak value in the period between pulses. To ensure this operation, the peak detector must be sufficiently phase-compensated for no overshoot and to keep the peak detector from following the input signal before the comparator can be

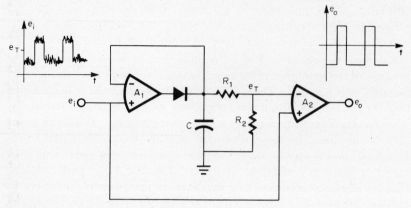

Fig. 3.2 Comparator pulse detection with automatically adjusted noise margin is achieved with the addition of a simple peak detector.

tripped. If there were no delay in the peak detector, the comparator threshold would increase at the same time that a pulse appeared to trigger the comparator.

Another characteristic commonly provided by comparators is hysteresis. Voltage comparators provide highly precise level detection when hysteresis is not required. However, the addition of hysteresis feedback generally degrades this precision. By switching a hysteresis signal into a comparator input instead of using hysteresis feedback, the inherent trip-point accuracy of the comparator can be retained.[3]

Comparator hysteresis is commonly produced by positive feedback from output to noninverting input.[1] The associated feedback voltage is not well controlled since it is derived from the output saturation voltages of the comparator. Greater accuracy is achieved by clamping the comparator output with zener diodes, but level detection remains limited by the zener tolerances and drifts. These zener voltage errors are much greater than the comparator input offset voltage and drift which otherwise limit level detection accuracy. The hysteresis errors are notably troublesome if signal zero crossing is to trigger the comparator. Hysteresis feedback splits the initial comparator trip point in two and shifts the resulting trip points away from zero. To shift one trip point back to zero, an added bias is required, and this bias must track the hysteresis feedback voltage to hold the trip point at zero.

To avoid the hysteresis feedback error and realize the inherent comparator accuracy, the comparator of Fig. 3.3 can be used. With this circuit there is independent voltage control of the first trip point and of the hysteresis that produces the second trip point. Note that the trip-point and hysteresis control voltages e_t and e_h can be made varying signals or constant depending on the control function desired. For the circuit shown

Signal Analyzers 61

hysteresis is produced by switching an operational amplifier input between ground and e_h. Switching is performed by the FETs Q_1 and Q_2 as the amplifier output changes states. Negligible error is introduced by the FET ON resistances because the only current they conduct is the input bias current of the amplifier and the leakages of the FETs and diodes. Such currents are readily controlled so that they develop voltages of only approximately 10 μV with the FET ON resistances.

When the amplifier output is in its negative state, Q_2 is held off, and Q_1 is on to connect the noninverting input to ground. Output switching will occur when the input signal e_i drives the inverting amplifier input to zero. This first trip point is at $e_i = -R_1 e_t/R_2$. To set this trip point at zero input voltage for zero-crossing detection, e_i is connected directly to the inverting input without e_t or the summing resistors R_1 and R_2. At the first trip point, the amplifier output switches positive to hold Q_1 off and let Q_2 turn on. Then, Q_2 connects the noninverting input to the hysteresis control voltage e_h. This establishes a second trip point at that level of e_i for which the inverting input is also at a voltage equal to e_h. Switching occurs at

$$e_i = -\frac{R_1}{R_2} e_t + \left(1 + \frac{R_1}{R_2}\right) e_h$$

for a hysteresis of $(1 + R_1/R_2) e_h$.

Trip-point errors are now only those introduced by the amplifier input errors and the summing resistors. Both trip points are offset by the dc errors of the amplifier input offset voltage and the flow of amplifier input

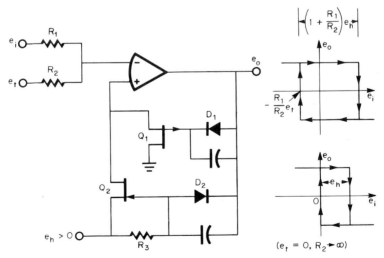

Fig. 3.3 Precise, independent control of comparator trip point and hysteresis is achieved by switching the hysteresis control signal to the comparator.

bias current in R_1 and R_2. Amplifier gain error shifts the trip points apart by a voltage equal to the peak-to-peak output swing divided by the amplifier gain.

Voltage comparators are also commonly used to provide an indication when a voltage monitored exceeds a reference level. Visual indication is readily provided by a lamp driven from the comparator, but audio indication requires addition of circuitry such as a keyed oscillator to drive a speaker or buzzer. To simplify this added circuitry, the comparator and oscillator functions can be made to time-share one operational amplifier[4] as shown in Fig. 3.4.

This circuit switches roles from comparator to oscillator when the input signal exceeds a reference level. As will be described, comparator operation with a controllable trip point results from the positive feedback through the R_1-nR_1 network and from the reference voltages developed by the zener diode. Oscillator operation is provided by the same elements in conjunction with the R_3-C feedback network.[4,5] Free-running oscillation in the low signal state is prevented by D_1, and the FET provides capacitor discharge.

When the input signal e_i is low, the amplifier output voltage is negative, reverse-biasing D_1. This blocks current to R_3, leaving the FET with zero gate-source bias. As a result, the FET is on and shunts the capacitor to hold the inverting amplifier input and e_{o1} at zero voltage. With that input effectively grounded, the amplifier is in a common comparator configuration.[1] Comparator switching will occur when e_i is large enough to raise the noninverting input above zero voltage, where the other input is clamped. The first comparator trip point is at nV_Z.

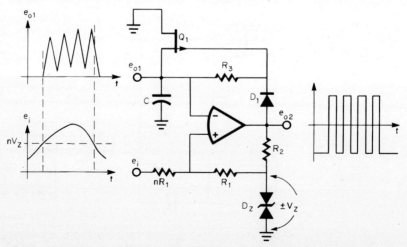

Fig. 3.4 A comparator generates an audio alarm signal when a high input signal switches the circuit to an oscillator.

When e_i reaches this trip point, the circuit switches to an oscillator mode. The amplifier output swings positive, forward-biasing D_1, to supply current to R_3, turn off the FET, and charge the capacitor C. Charging continues until the hysteresis-produced second trip point is reached. That trip point is reached when the capacitor voltage is at $(e_i + nV_Z)/(n + 1)$.

When the capacitor voltage reaches the second trip point, the amplifier output swings negative to again reverse-bias D_1 and turn on the FET. The FET discharges C until the capacitor voltage reaches zero or some greater voltage maintained by e_i at the noninverting amplifier input. That input will be above zero voltage as long as e_i remains above the first trip level of nV_Z. In that case discharging is interrupted before reaching zero, and the circuit switches back to the charging mode. This charging and discharging oscillation will continue as long as e_i remains above the first trip point. If R_3 and the FET are chosen for similar charging and discharging currents, a roughly triangular voltage waveform will be produced on C for an audio output signal e_{o1}. When e_i drops below the first trip point, the voltage at the noninverting input will be below zero. The capacitor cannot be discharged below zero voltage by the FET to again switch the amplifier state; so oscillation will then stop.

Circuit performance is precise for the comparator function. The key to circuit performance is the first trip point because this determines the oscillator turn-on. Normal comparator errors affect this trip point, and they are primarily due to the tolerance and thermal variations of the zener diode voltage. Tolerance error can be compensated for by adjustment of the R_1 and nR_1 resistors.

Precise control of the oscillator waveform is not required for audio alarm requirements. The oscillator waveform will vary because of the presence of the input signal, an imprecise capacitor discharge current, and any loading current drawn from the capacitor. As mentioned above, the presence of e_i shifts the circuit switching points for oscillator operation. From these shifts the signal e_{o1} is offset by a voltage equal to $e_i/(n + 1)$, and the charging current is reduced. Oscillation frequency changes with charging current variation. Frequency also varies with the discharge current supplied by the FET and with any loading current diverted from the capacitor. Where this loading is severe, a simple emitter-follower buffer can be added. Frequency can be controlled within an acceptable 30 percent range of design center, which is roughly

$$f \approx \frac{1}{2\pi \left(R_3 C \ln \frac{E_{0+}}{E_{0+} - V_Z} + \frac{CV_Z}{I_{DSS}} \right)}$$

where E_{0+} is the positive saturation voltage of the amplifier.

3.1.2 Window comparators

A window comparator provides an indication of whether a signal is within a given voltage range. If a signal within the defined range produces a high comparator output state, then a low output state will result for signals that are either above or below that range. Such comparator action is commonly required for go/no-go testing that checks for operation within a given range of performance.

Window comparator operation is illustrated by the transfer characteristic of Fig. 3.5. As shown, a window comparator typically consists of one operational amplifier, or comparator, that detects signals above the acceptance range and another to detect signals below the range. The two amplifiers compare the signal against separate references. To provide a single output signal, the amplifier output signals are combined by a NAND gate. Only when both amplifier outputs are low will the gate output be high. For this to occur, the input signal e_i must be greater than reference voltage E_1 and less than E_2. Comparison accuracy is largely determined by the amplifier input offset voltages and gain errors plus the parasitic hysteresis described in the next subsection.

A slightly simpler window comparator configuration is available, and it is readily expanded to multiple window operation. As shown in Fig. 3.6, this configuration makes use of an FET switch instead of an NAND gate, and the output is taken from only one amplifier. Window comparator action results from switching the input signal path of A_1 when the threshold A_2 is reached. If the input signal e_i is lower than the E_2 threshold, the output of A_2 will be negative; so the FET is on. This connects e_i to one input of A_1, where it is compared against reference E_1. If e_i is lower than E_1, the output e_{o1} is negative. When e_i increases to the level of E_1, e_{o1} switches to its positive state and remains there with increasing e_i until the level E_2 is reached. Then comparator A_2 switches to its positive state and turns off the FET switch. This disconnects the input signal from A_1, allowing the resistor R to drop the noninverting input of A_1 to the negative supply level. As a result, e_{o1} is driven back to its negative level.

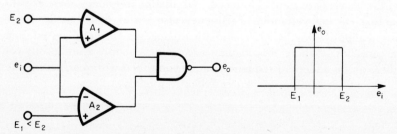

Fig. 3.5 Common window comparators are formed with two operational amplifiers or comparators that compare the signal against different reference levels plus an NAND gate for combining the amplifier output signals.

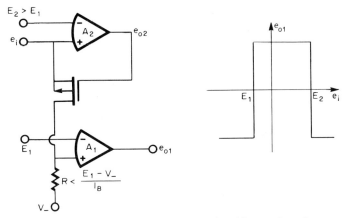

Fig. 3.6 Window comparator action is also produced by switching the input signal away from a comparator when a second threshold is reached.

Additional comparison error results with this configuration because of the switch. In addition to the input offset voltage gain and hysteresis errors of the last circuit, the circuit of Fig. 3.6 encounters a comparison error due to switch ON resistance and a transient error associated with switch capacitance. Very small ON resistance error can be maintained by making R large to limit switch current. That resistor must be small enough to conduct the input bias current of A_1 when the switch is off, but this still permits switch currents as low as the amplifier input current. Such low currents, however, take considerable time to discharge switch capacitance during switching transitions. Thus, R must be chosen considering the compromise between comparison accuracy and switching time.

An extension of the circuit of Fig. 3.6 provides multiple windows for amplitude classification or sorting. Test sorting separates things into various ranges of some characteristic, and these ranges can be electrically defined by separate, adjacent comparator windows. Such an arrangement of electronic windows is provided by the circuit of Fig. 3.7. Each window is defined by two adjacent comparators in the same manner as described with the previous circuit. Together A_1 and A_2 define a window between E_1 and E_2 in the output response of A_1, and similarly for A_2 and A_3, A_3 and A_4, etc.

As the input signal rises through the reference levels of the various comparators, their outputs are successively driven positive. Along with this, the associated signal-conducting FETs are turned off to return the previous comparator output to its negative state. Only the last, or nth, comparator output is not switched back to its negative state. Otherwise each comparator output remains positive only while e_i is at a level between

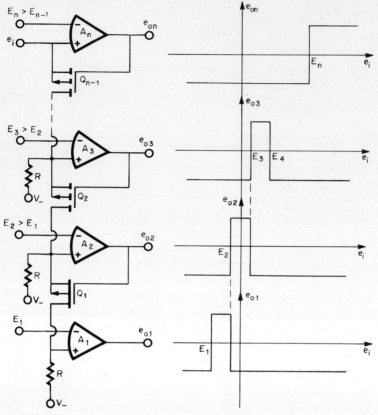

Fig. 3.7 Multiple, adjacent comparator windows for sorting or amplitude classification are developed by extending the circuit of Fig. 3.6.

the reference voltage of that comparator and that of the next higher. As with the preceding circuit, the switch ON resistances and capacitances introduce comparison and response errors that place conflicting demands on the resistance chosen for R. This compromise can be avoided with greater circuit complexity by simply using multiple comparators with their outputs combined by digital circuitry.

Basic window comparator operation can also be attained with only one operational amplifier or comparator, where comparison accuracy is less critical. In the simplest case, window comparison is achieved by diode-gating the input signal to the two inputs of an amplifier, as in Fig. 3.8. Negative input signals that forward-bias D_1 will force the amplifier output positive, as will positive signals that forward-bias D_2. Because opposite polarity signals are gated to opposite amplifier inputs, they produce the same output polarity. However, signals too low to forward-bias either diode will not determine the output state. For this intermediate condition, a bias resistor R_2 is added to ensure a negative output in the absence of signal at the amplifier inputs. Alternatively, a negative output in the window can be developed through the use of the amplifier offset voltage null control. In either case, the offset supplied must be large enough to ensure switching at a rate compatible with the input signal frequency.

Significant error affects this simple window comparator since the diode voltages serve as the references defining the window. The diode voltages are subject to significant tolerance and thermal variations. Better accuracy would result with zener diodes and their controlled voltage drops. However, this circuit would not realize the inherent zener diode accuracy since diode currents would vary directly with the signal. As zener diode current increases from zero, the zener voltage undergoes significant change, and this turn-on characteristic is temperature-sensitive. Instead, it is better to maintain continuous zener bias with the window comparator of Fig. 3.9.[6] One zener diode now serves to define both the upper and lower band limits; so these limits will be well matched about zero. For limits not centered about zero, the range center can be shifted by connecting bias resistors to the amplifier inputs from the appropriate power-supply voltage.

Window comparator operation with this circuit also results from diode gating of the input signal to the appropriate amplifier input. Input signals above the positive limit forward-bias D_1 and pull the zener upward in voltage to forward-bias D_4. This applies a positive voltage to the noninverting input of the amplifier, causing the output to swing to its positive state. For this to happen, the input signal must support the zener voltage and two forward diode voltages so that the upper range limit is $V_Z + 2V_F$. A positive output swing is also produced by negative input signals that reach $-V_Z - 2V_F$. Such signals forward-bias D_2 and D_3 to apply a negative

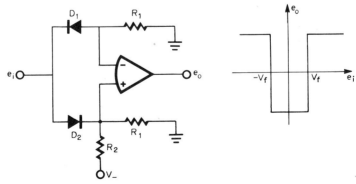

Fig. 3.8 A simplified window comparator diode gates opposite polarity signals to opposite amplifier inputs and uses the diode voltages to define the window.

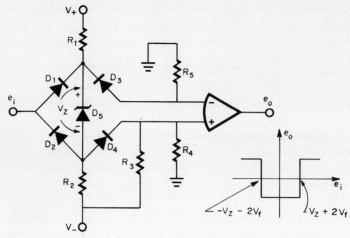

Fig. 3.9 Window comparator operation is achieved with a single operational amplifier and a diode bridge that gates high and low signals to opposite amplifier inputs when the signal magnitudes exceed the reference level.

signal to the inverting amplifier input. Signals within the band defined by the above limits are not conducted to the amplifier, and a negative output state is produced by the bias from R_3.

Comparison accuracy is controlled by diode voltages at low frequencies and by amplifier gain bandwidth at higher frequencies. The zener and junction diode voltages defining the comparison band are subject to tolerance and temperature variations. As a result, the band limits are typically in error by several percent. To reduce the temperature sensitivities of the band limits, R_1 and R_2 are selected to bias the zener diode so that its thermal voltage variation approximately cancels those of two junction diodes. Adding to the dc error is the offset shift introduced by R_3; so this shift is kept comparatively small.

At higher frequencies the comparator error is dominated by the gain-bandwidth-limited output swing from the positive to the negative state. This transition occurs when the input signal is disconnected from the amplifier by the diodes, leaving at the amplifier input only the small voltage developed by R_3. Amplifier input voltage larger than this is required to cause switching at some higher frequency. This frequency limit for operation and related delay errors at lower frequencies are determined by the gain-bandwidth product of the operational amplifier. Gain-bandwidth product is improved by removing phase compensation from operational amplifiers used in this comparator.

Other response limitations to consider with this circuit are the amplifier overload recovery delay and the diode capacitance discharging. In order to switch, the amplifier must first recover from its saturated condition; so a time delay is introduced. Fortunately, the removal of phase compensation ensures short overload recovery for most operational amplifiers. Another switching delay can be produced by the discharging of the capacitances of D_3 and D_4 through R_4 and R_5. This factor and circuit input resistance govern the choice of R_4 and R_5. Input resistance equals one of these resistors shunted by R_1 and R_2.

3.1.3 Reducing comparator hysteresis[7]

Deceptively large error is introduced into comparator operation by parasitic hysteresis. This hysteresis results from nonzero switching time and restricts accurate voltage comparison to far slower signals than would be anticipated from comparator bandwidth and slewing-rate specifications. For example, a comparator having a 100 V/μs slewing rate can swing a full ± 10 V output at over 1 MHz rate. But input signals that even approach this frequency develop very large comparison errors. At only 250 kHz, comparison error reaches 10 percent of the signal magnitude for the more easily handled triangle- and sine-wave signals. Even greater error results with more rapidly changing signals, such as pulses or square waves. The source of this parasitic hysteresis error and means for reducing it are described below.

Gain error and switching time cause a comparator response to deviate from the ideal as represented in Fig. 3.10. Ideally the output would switch exactly at the zero input point, as in Fig. 3.10a. However, the comparator gain is finite, and small input signals are not amplified enough to force the comparator output to one of its limits E_{op} or $-E_{op}$. As the input signal e_i passes through this range of small voltages, the output voltage e_o follows e_i linearly with a slope of $-1/A$ in Fig. 3.10b. Thus, e_i must

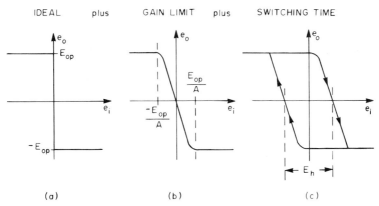

Fig. 3.10 Comparator input-output response differs from the ideal by a noninfinite gain slope and a hysteresis resulting from nonzero switching time.

rise above E_{op}/A or drop below $-E_{op}/A$ to complete the comparator output swing. Since the comparator gain A decreases with increasing frequency, this region of linear response increases with the signal frequency.

Also altering the comparator response from the ideal is the parasitic hysteresis represented in Fig. 3.10c. This hysteresis is a major source of comparator error and is created by the nonzero switching time of a comparator. To show how this hysteresis results, the coincidence of comparator input and output signals is plotted in Fig. 3.11. Ideally, the output signal e_o would pass through zero at the same time as the input signal e_i. However, e_o must first swing from its peak level E_{op} or $-E_{op}$, and the time required for this swing is a hysteresis-inducing delay t_{h-} or t_{h+}. During this time, the input signal moves away from zero, and the level it reaches before e_o crosses zero appears as a comparison error. Opposite polarity errors result for rising and falling e_i; so a separate path is developed for each case on the input-output transfer characteristic shown. The difference in e_i for the two e_o zero crossings equals the response hysteresis E_h.

The greater the rate of change of e_i, the more it will be away from zero before e_o reaches zero; so the greater the hysteresis. Hysteresis is related to the rate of change in the input signal rather than its frequency. This relationship can be expressed in simple form to permit determination of the comparator characteristics needed to limit hysteresis error to a desired level. An exact solution for the hysteresis error would require a mathematical expression of the input waveform and would result in a different expression for each type of waveform. Instead, a more general expression can be derived if it is assumed that the rate of change of the input

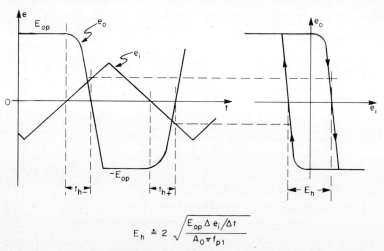

$$E_h \doteq 2\sqrt{\frac{E_{op}\, \Delta e_i/\Delta t}{A_0 \pi f_{p1}}}$$

Fig. 3.11 Comparator parasitic hysteresis results from the delays t_{h-} and t_{h+} between input and output zero crossings as produced by limited gain-bandwidth product.

signal remains constant during the comparator switching transition. General expressions are then found for the cases of rise-time-limited switching and slew-rate-limited switching.

For the rise-time-limited case, a far simpler expression results if it is assumed that the comparator has a single-pole frequency response and switching time small compared with the time constant of this pole. Then, the output switching time to zero is

$$t_{h-} = t_{h+} \doteq \sqrt{\frac{E_{op}}{A_0 \pi f_{p1} \Delta e_i/\Delta t}}$$

where A_0 is the dc gain and f_{p1} is the first response pole of the comparator. The hysteresis voltage developed by these switching times equals the change in e_i during t_{h-} and t_{h+}:

$$E_h \doteq 2 t_{h-} \frac{\Delta e_i}{\Delta t} \doteq 2\sqrt{\frac{E_{op} \Delta e_i/\Delta t}{A_0 \pi f_{p1}}}$$

While the accuracy of this expression is restricted by the various assumptions made, it does provide an order-of-magnitude indication of E_h.

A simpler, more accurate, expression results for the case of the slew-rate-limited comparator output of Fig. 3.12. In this case the average slope of the output transitions equals the slewing rate S_r. The hysteresis delay times t_{h-} and t_{h+} are then merely E_{op} and $-E_{op}$ divided by the slope S_r. Thus the hysteresis is

$$E_h = \frac{\Delta e_i}{\Delta t}(t_{h-} + t_{h+}) = \frac{\Delta e_i}{\Delta t} \cdot \frac{E_{opp}}{S_r}$$

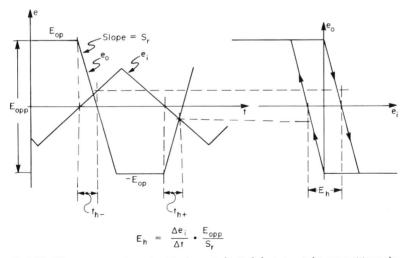

Fig. 3.12 When a comparator output is slew-rate-limited during switching transitions, the parasitic hysteresis is inversely related to the slewing rate S_r.

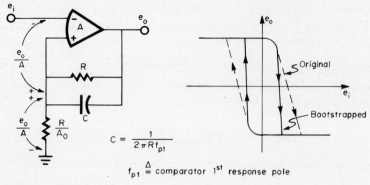

Fig. 3.13 Where signals are too slow to drive a comparator into rate limiting, positive feedback can be used to bootstrap the input and reduce hysteresis if the feedback factor equals the inverse of the comparator gain.

for e_o rate-limited and $\Delta e_i/\Delta t$ constant during the switching transitions. Note from the two expressions for E_h that this parasitic hysteresis is related to the input signal rate of change and that higher comparator gain-bandwidth product and slewing rate are needed to reduce E_h.

Hysteresis can also be reduced by several circuit techniques including limited bootstrapping, output swing limiting, and phase-lead compensation of the switching point. Bootstrapping or positive feedback is commonly applied to add hysteresis to comparator response, but an appropriate amount will reduce hysteresis where the comparator output is not rate-limited. It does so because of the switching speed improvement also provided by positive feedback. The positive feedback factor appropriate for hysteresis reduction is equal to the inverse of the comparator open-loop gain as provided in Fig. 3.13. To make this feedback track the comparator gain as frequency increases, R is bypassed by a capacitance

$$C = \frac{1}{2\pi R f_{p1}}$$

where f_{p1} is the frequency of the comparator first response pole. Resistor R should be made large to limit C to a capacitance level that does not slow the comparator switching.

Hysteresis reduction by positive feedback is limited by the comparator rise time and slewing rate. Until the comparator output rises somewhat, there is little feedback signal to speed switching. The initial output rise is delayed by phase shift in the comparator; so much of the switching time and its associated hysteresis remains. Once the positive feedback signal is developed, switching continues at a rate up to the slew-rate limit of the comparator.

Where the comparator is driven to its switching-rate limit by the input signal alone, hysteresis can be reduced by simply clamping the output swing. For many applications a 0 to 5 V output swing is adequate. Reduced output voltage swing results in a proportional decrease in switching time if the swing is rate-limited rather than rise-time-limited. Much less improvement is achieved for the rise-time-limited case because the slow, initial portion of the output rise remains.

For hysteresis reduction, output clamping to ground is often not suitable because it overloads the comparator output, producing overload recovery delay. Instead, feedback clamping can be used to avoid the overload. The type of feedback clamp appropriate depends upon whether the comparator is frequency-stable under negative feedback. Such stability is assured with phase-compensated operational amplifiers operated as comparators, and clamping can be performed with a single zener diode as in Fig. 3.14. Positive output swings are limited to the zener voltage, and negative swings are clamped by the forward biasing of the same diode. The reduced output swing on the transfer plot of Fig. 3.14 is accompanied by a decreased hysteresis.

The simple feedback clamping above is not suitable for many comparators since they are not frequency-stable with negative feedback. Even when operational amplifiers are used as comparators, it is desirable to remove the stabilizing but speed-limiting phase compensation. In these cases a more specialized feedback clamp can be used. That clamp permits switching unencumbered by phase compensation, but then provides the stabilizing compensation in the clamped states. Such a clamp is shown in Fig. 3.15 for operational amplifiers that have provisions for external feedback phase compensation.

When the output is in its positive state, D_1 forward-biases to turn on D_Z for clamping at $V_Z + V_{f1}$. Also, D_1 connects the output to C_1 for phase-compensating feedback. Similarly, for the negative output, D_2 conducts to provide clamping at $-V_{f2} - V_{f3}$ and to connect the output to compensation capacitor C_2. But during the switching transition both D_1 and D_2

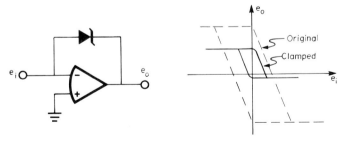

Fig. 3.14 Clamping output swing greatly reduces the time required for comparator switching and the associated parasitic hysteresis as long as e_i is fast enough to ensure rate-limited switching.

are reverse-biased so that neither phase compensation capacitor is driven by the output. Effectively, the phase compensation is disconnected during the switching so that it does not introduce delay. Instead, switching time and its associated hysteresis are decreased by the reduction in output swing.

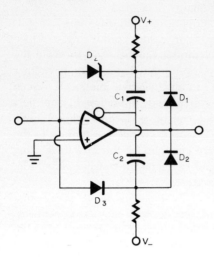

Fig. 3.15 Comparator frequency stability under feedback clamping without phase-compensation-limited switching speed is provided by a clamp circuit that connects the compensation only in the clamped states.

One further means of reducing comparator parasitic hysteresis is phase-lead compensation of switching points. If the phase lead is set to counteract the phase lag of the comparator, hysteresis is reduced to zero. The desired phase lead can be developed by a differentiator to compensate the hysteresis described earlier by

$$E_h \doteq 2 \sqrt{\frac{E_{op} \, \Delta e_i/\Delta t}{A_0 \, \pi \, f_{p1}}} \quad \text{for rise-time-limited switching}$$

$$E_h = \frac{\Delta e_i}{\Delta t} \cdot \frac{E_{opp}}{S_r} \quad \text{for slew-rate-limited switching}$$

Note that in both expressions the hysteresis is related to the rate of change or time derivative of e_i during the switching transition. Thus, the signal produced from e_i by a differentiator can be used to compensate E_h. For the rise-time-limited case the compensation is approximate since E_h is proportional to the square root of $\Delta e_i/\Delta t$.

This phase-lead compensation is provided by simply connecting a differentiator between the signal input and the noninverting comparator input as in Fig. 3.16. With this connection the differentiator controls the comparator trip point so that switching begins before the input signal zero crossing. Waveforms illustrating this operation are presented in Fig. 3.17. An input signal e_i approaching zero with a positive slope

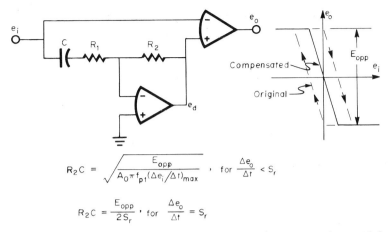

Fig. 3.16 Phase-lead compensation of comparator parasitic hysteresis can be provided by a differentiator if the input signal slope is constant during the switching transitions.

produces a negative differentiator output voltage e_d. That voltage causes the output voltage e_o to begin its switching transition before e_i reaches zero. By the time e_i reaches zero, so will e_o, for an appropriate shift in trip point by e_d. Similarly, an input signal of negative slope results in a positive e_d to again initiate comparator switching prior to the e_i zero crossing. If e_d again results in simultaneous zero crossings for e_o and e_i, hysteresis is reduced to zero.

To achieve this hysteresis compensation, the differentiator gain R_2C is selected for the $\Delta e_i/\Delta t$ range of interest. If $\Delta e_i/\Delta t$ is large enough to

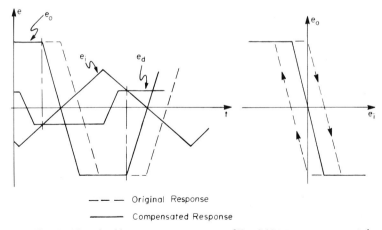

Fig. 3.17 The phase-lead hysteresis compensation of Fig. 3.16 initiates output switching prior to the input signal zero crossing so that input and output signals cross zero simultaneously.

force the comparator into rate-limited switching, then E_h from above is reduced to zero for $R_2C = E_{opp}/2\,S_r$. For smaller $\Delta e_i/\Delta t$, where this phase-lead compensation is not exact, minimum error is achieved for a differentiator gain that makes hysteresis zero near the maximum $\Delta e_i/\Delta t$ anticipated. Then,

$$R_2C = \sqrt{\frac{E_{opp}}{A_O\,\pi\,f_{p1}\,(\Delta e_i/\Delta t)_{max}}}$$

Compensation is limited to a maximum $\Delta e_i/\Delta t$ by the response of the differentiator. Fortunately, less stringent response requirements are placed on the differentiator than on the comparator. While the comparator would have to swing its full output range in zero time for zero hysteresis, the differentiator output need only settle to the smaller compensation voltage level before the input signal reaches that level. Thus, the more restricted bandwidth of the differentiator is compatible with a comparator whose speed is not limited by phase compensation.

3.2 Peak Detectors[8,9]

The most common measure of an ac signal voltage or current is its rms value, because it provides a meaningful indication of energy content; but rms measuring instruments cannot measure a transient signal, such as a single pulse. Such pulses are encountered in spectral analyses as with gas chromatographs and mass spectrometers. Nor does the rms instrument convey signal amplitude information unless the signal is repetitive and of a mathematically described waveform. Signals do not meet these conditions in peak-to-peak noise measurement or in process control deviation monitoring. Even when the signal does meet these conditions, amplitude is sometimes of more interest than rms value. In all these signal measuring requirements a peak detector replaces the rms meter. Described here are basic and specialized peak detectors applicable to common measurement needs. Techniques for greatly improving peak detector accuracy and speed are also presented.

3.2.1 Basic peak detectors A peak detector is essentially a sample-hold circuit whose operating modes are controlled by the signal monitored. As the signal rises above a previously held voltage, a peak detector follows the signal in its SAMPLE mode. When the signal begins to decrease, the peak detector holds the maximum previous voltage of the signal in a HOLD mode. This voltage level is maintained at the peak detector output until forced higher by a larger input signal or until reset to zero. By this operation, a peak detector provides a dc output equal to the maximum amplitude of the signal monitored.

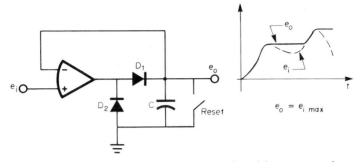

Fig. 3.18 To store peak signal levels, an operational amplifier is connected as a voltage follower that can supply charging but not discharging current to a capacitor.

Peak detector operation is illustrated in Fig. 3.18 along with a basic peak detector circuit.[1] Because of the gating of the diodes shown, the detector output can be increased but not decreased by the input signal. If e_i rises above a previously stored voltage, the operational amplifier senses this difference and responds with a positive output swing. This forward-biases D_1 to connect the amplifier as a voltage follower driving the holding capacitor, and the capacitor voltage follows the input signal. When e_i decreases, the amplifier output swings negative, but D_1 blocks capacitor discharge current. The capacitor voltage then remains at the maximum previous input level.

To remove this stored voltage before a new measurement cycle, the RESET switch shown can be used. This switch is most commonly a relay or an FET. Although the FET is simpler, it introduces leakage and offset errors. These errors are described later along with techniques for removing them. An alternative to the RESET switch is a shunt resistor if the signal is repetitive. A resistor can be chosen that will not produce significant discharge between repetitive signal peaks but that will eventually discharge the capacitor in absence of signal.

Other errors of the above peak detector are due to the normal voltage-follower errors of the operational amplifier, overshoot, diode switching time, and parasitic drains on the holding capacitor C. SAMPLE mode errors are introduced by the amplifier input offset voltage, gain error, common-mode error, and bandwidth in the same way as with the common voltage follower. In addition to these errors, amplifier overshoot can charge the capacitor to a voltage greater than the peak. To avoid this potentially large error, the amplifier must be critically damped or overdamped by phase compensation. Normally phase-compensated operational amplifiers are not likely to be suitable, because they are often underdamped, and the loading of the storage capacitor creates more overshoot. As a result, it is generally necessary to use operational amplifiers having provision for external phase compensation. To select phase compensation, it is desirable to observe the square-wave response of the circuit.[1] However, the peak detector would respond only to the first peak of the square wave; so the detector diode is temporarily shorted to provide repetitive response signals. With large storage capacitors, phase compensation is eased by a small decoupling resistor in series with the amplifier output.

HOLD mode error is created by parasitic drains on C which cause the output to change with time or to droop. Droop-inducing parasitics include the amplifier input current, leakages of D_1 and the RESET switch, dielectric absorption of the capacitor, and any output load current. To reduce the amplifier input current, an FET input type can be used. In any case, an operational amplifier with high input impedance under overload should be used. Under overload conditions such as that of the HOLD mode, many operational amplifiers have low input impedance associated with input protection clamps. Leakage-induced droop is reduced by the careful choice of diode D_1 and the RESET switch, as well as by means of leakage decoupling techniques to be described. Dielectric absorption is minimized by using Teflon or polystyrene capacitors.

Error can also result from the time required to switch from the HOLD mode to the SAMPLE mode. During this transition, the amplifier output must swing from where it is clamped by D_2 to a diode drop above the input signal. Only then can the circuit acquire a new signal peak. If this transition is longer than the duration of a peak, that peak will be missed. This transition time is the major limitation to acquisition time. Some improvement is achieved by clamping the output with D_2 to reduce the voltage span of the transition. Other more effective techniques for reducing transition time are described later.

Peak detector design begins with a definition of the permissible droop and acquisition time. For a given droop rate D caused by a net parasitic drain current I_P, the holding capacitor must be

$$C_H \geq \frac{I_P}{D}$$

To charge this capacitor within an acquisition time t_a, an operational amplifier must be chosen for compatible settling time and output current. The amplifier settling time must be no longer than t_a when the amplifier is phase-compensated for zero overshoot with a capacitive load equal to C_H. And the amplifier output current I_O has to charge the holding capacitor to a peak voltage E_P in less than the acquisition time. This requires an output current of typically

$$I_O \geq 2C_H \frac{E_P}{t_a}$$

In this expression the factor of 2 arbitrarily allots one-half the acquisition time for charging C_H and the other half for final settling.

Most loads connected to the output of the above peak detector would cause significant droop because the capacitor would supply the load current in the HOLD mode. Thus, output buffering is commonly included. A simple buffering is provided by adding two FETs to the previous circuit as in Fig. 3.19. This replaces the load current drain on the capacitor with the small gate leakage of Q_1. It also removes the drain of amplifier input current from the capacitor; so a bipolar input rather than FET input operational amplifier can be used. No SAMPLE mode error is added by the offset voltage and output resistance of this buffer. When in the SAMPLE mode, feedback adjusts the capacitor voltage to counteract the buffer errors. However, in the HOLD mode this error-correcting feedback is removed, and any change in buffer offset or loading will create error. Accuracy is then dependent on stable FET temperatures and constant loading. The output resistance is $R_S + 1/g_{fs}$.

Some temperature variation can be tolerated because the bias shown is temperature-compensated. Although the buffer is essentially a source-follower circuit, low offset and drift result from biasing the source follower with a matching FET current source to cancel the gate-source voltage shift of Q_1. The source resistor of Q_1 has a voltage drop set by the current of Q_2 to nearly equal the gate-source voltage of Q_2 if output current is small. If the two FETs have the same source current level, the gate-source voltage of Q_2 is nearly equal to that of Q_1. Thus the offset shift and its drift nearly cancel if output currents are small.

Alternatively, better output buffering is achieved with a voltage-follower-connected operational amplifier. Direct connection of a voltage follower to the circuit of Fig. 3.18 does, of course, add another set of offset voltage,

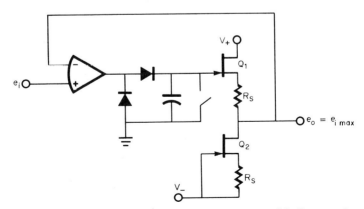

Fig. 3.19 A pair of FETs provide temperature-compensated buffering to the output of the peak detector of Fig. 3.18.

input bias current, and gain errors to the peak detector. To avoid this added error, the follower is connected within the feedback loop just as the previous buffer was. This results in the common peak detector circuit of Fig. 3.20. Since the buffer is within the feedback loop, its offset and gain errors are removed. Again the input current of the input amplifier is removed from the storage capacitor; so still only one amplifier input current loads the capacitor. Thus, peak detector error is not increased by the dc errors of an added voltage-follower buffer if that buffer is connected within the SAMPLE mode feedback loop.

To ensure that buffer ac error does not increase peak detector error, additional care must be taken with phase compensation. Phase compensation and operational amplifiers must be selected to ensure that the buffer A_2 is significantly faster than the input amplifier A_1. If A_1 could charge the capacitor faster than A_2 could follow, a feedback delay would be introduced that would allow overcharging of the capacitor. Such error is avoided by selecting phase compensation for A_1 that eliminates the peak detector overshoot. As before, this selection is made with a square-wave input signal and with the detector diode D_1 shorted to permit repetitive circuit response observation.

Also provided by the configuration of Fig. 3.20 is a peak detector with gain greater than unity. Gain is derived through the addition of resistor R_1. This provides an input-to-output response of a noninverting amplifier in the SAMPLE mode. Since the capacitor voltage equals the output voltage, the voltage stored will be an amplified equivalent of the peak input level.

To detect the magnitude of negative peaks or the minima of bipolar signals, the same circuits described above can be used if the diodes are reversed. Or with the circuit of Fig. 3.20, the input signal can be connected to R_1 rather than the noninverting input of A_1. Since this connects

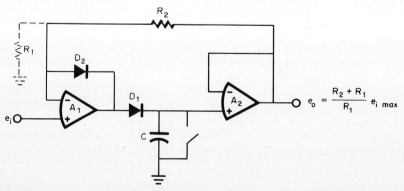

Fig. 3.20 Improved output buffering and peak detector gain greater than unity are achieved with an output voltage follower.

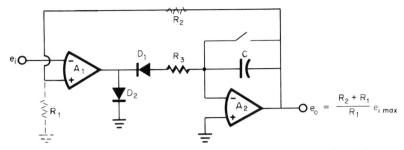

Fig. 3.21 The lower input bias currents of some inverting-only operational amplifiers can reduce droop if feedback is rearranged to permit an inverting output buffer.

A_1 as an inverting amplifier, it also makes possible the use of many inverting-only operational amplifiers which have higher speed. This response speed is needed to overcome switching time limitations faced by A_1. If the diodes are reversed along with this change in input signal connection, the circuit again detects positive peaks, and an inverting-only, fast operational amplifier can still be used.

To minimize droop caused by the buffer amplifier input bias current, it is sometimes desirable to use an inverting-only operational amplifier for the buffer. This makes possible the use of many inverting-only chopper-stabilized and varactor amplifiers which have very low input bias currents and associated thermal drifts. Such amplifiers can be used by connecting the holding capacitor in the feedback path of the output buffer as in Fig. 3.21. Circuit operation is much like that of the circuit of Fig. 3.20, and a detector gain greater than unity can be attained by adding R_1 and R_2. When the input signal rises above that fed back at the junction of R_1 and R_2, the output of A_1 swings negative. This forward-biases D_1, permitting charging of C, which raises the output voltage until the signal fed back matches the input signal. When the input signal begins to decrease, the output of A_1 swings positive and is clamped by D_2. Since D_1 disconnects C from positive signals, C holds the output voltage at the highest level previously reached. Note that feedback is connected to the noninverting rather than the inverting input of A_1 because there is now a phase inversion through A_2. To prevent delay-inducing overload at the input of A_2, resistor R_3 may be required.

3.2.2 Specialized peak detectors Two more specialized peak detectors are a peak-to-peak detector and a peak magnitude detector. A peak-to-peak detector is directly suited to measurement of bipolar, nonrepetitive signals, such as noise. It is also well suited to amplitude measurement of ac signals having dc offsets. As long as the offset is smaller than the signal, a peak-to-peak detector rejects the offset component without resorting to coupling capacitors. This avoids the measurement delay normally imposed by coupling capacitor charging. That delay is often the speed-limiting factor in automated testing. A peak-to-peak detector could be formed with separate positive and negative peak detectors built similar to Fig. 3.20, and this would require four operational amplifiers. However, the output would have to be measured differentially between the separate detector outputs, unless ground reference were restored with a fifth amplifier.

A simpler peak-to-peak detector circuit provides a ground-referenced output, requires only three operational amplifiers, and has inherent droop compensation. This circuit is essentially a combination of the simple peak detector of Fig. 3.18 with that of Fig. 3.21, and the result is shown in Fig. 3.22. That portion of the circuit derived from Fig. 3.18 is the positive peak detector formed with A_2, D_1, D_2, and C_1. Although the input to this detector is halved by the R_1 divider, the output from C_1 receives an adjusting gain of 2 through A_3. In addition, A_3 combines with A_1 and their feedback elements to form an inverting, negative peak detector similar to Fig. 3.21. Note that the ground reference of this negative peak detector is here taken from the output of the positive peak detector. Because of this choice of reference, A_3 adds the amplified output of the positive peak detector to that of the inverting, negative peak detector. This sum is the peak-to-peak maximum of the input signal. It is to permit the referencing of one detector from the other that the R_1 divider is added.

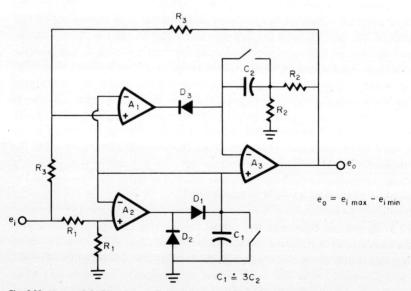

$$e_o = e_{i\,max} - e_{i\,min}$$

$$C_1 \doteq 3C_2$$

Fig. 3.22 A simplified peak-to-peak detector is formed with the simple peak detector of Fig. 3.18 and the peak detector of Fig. 3.21.

A significant accuracy improvement is inherent in this peak-to-peak detector configuration. Because there are holding capacitors connected to both inputs of A_3, its input bias currents create canceling droop effects. Also, by appropriate choice of capacitor values, the droop on C_1 caused by the input currents of A_1 and A_2 can be counteracted. If the three amplifiers have similar input bias currents, then the design center for droop compensation is $C_1 = 3C_2$. Even better compensation is attained by experimentally nulling droop through the addition of capacitance to either C_1 or C_2.

Otherwise, the accuracy limitations of this peak-to-peak detector are related to the errors of its individual peak detectors as described for Figs. 3.18 and 3.21. The errors of these basic peak detectors can be summed to find the error of the composite circuit. In making this summation the effects of the R_1 and R_2 divider networks must be considered. Because of the R_1 divider, the input offset voltage of A_2 will be twice as significant as in the basic circuit. Similarly, the R_2 divider multiplies the offset voltage of A_3 by 2, and the associated increase in closed-loop gain doubles the gain error of A_3. Operating speed is still primarily limited by mode switching times; but if R_2 is too large, its charging of C_2 will pose an acquisition time limit. As with previous circuits, A_1 and A_2 must retain high input impedance under overload to avoid excessive drain on C_1 when either or both of these amplifiers are in a HOLD mode. Also, both A_1 and A_2 must be phase-compensated for zero overshoot.

A common application of peak detectors is in the detection of the maximum excursion of a signal, such as the deviation of a process control monitor from a set point. Such deviations can often be either positive or negative, and both polarities cannot be monitored by one basic peak detector. Instead, two peak detectors could be used with a maxima selector that picks the larger detector output. A simpler means of detecting maximum signal deviation is provided by the magnitude peak detector of Fig. 3.23. This detector stores the peak of the signal magnitude, which is the peak of the absolute value of the signal.

Essentially the detector consists of a common positive peak detector and an inverting, negative peak detector combined to use the same storage capacitor and output buffer. Together A_1 and A_3 perform as a positive peak detector such as that described earlier in Fig. 3.20. Positive peaks above the voltage held by the capacitor will cause the output of A_1 to swing positive and increase the capacitor voltage to the higher input level. For negative signal excursions, A_2 and A_3 perform as an inverting, negative peak detector. If the magnitude of a negative excursion exceeds the voltage stored, the R_1 feedback network will drive the inverting input of A_2 negative. This causes the output of A_1 to swing positive to charge the capacitor until $e_o = -e_i$ and the inverting input of A_2 is near

84 Designing with Operational Amplifiers

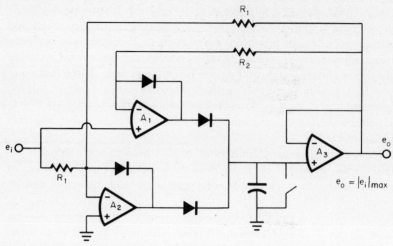

Fig. 3.23 A magnitude peak detector is formed by a simplified combination of a positive peak detector and an inverting, negative peak detector.

zero. In both positive and negative peak detector modes the holding capacitor is charged positive; so the detector output is independent of the polarities of the signal peaks. Accuracy and speed limitations include those of the detector of Fig. 3.20, those of a common inverter added by A_2, plus any overshoot error.

The above magnitude peak detector can be modified for gains greater than unity by altering both the positive detector and negative detector feedbacks. To increase the positive detector gain, a resistor is added to ground from the inverting input of A_1 as discussed with Fig. 3.20. A similar gain increase for the negative detector of Fig. 3.23 is created by making the R_1 feedback resistor greater than the R_1 input resistor, just as is done for an inverting operational amplifier for gain greater than unity.

3.2.3 Improving peak detector accuracy

The two major types of peak detector error are droop during HOLD and acquisition error upon switching to SAMPLE. Acquisition errors are primarily controlled by the response speed of the detector, and error-reducing speed improvements are discussed later. As previously mentioned, droop error results from parasitic current drains on the storage capacitor during the HOLD mode. Principal sources of these parasitic currents are the input bias currents of amplifiers, detector diode leakage, and RESET switch leakage. Techniques for reducing each are presented below along with means for avoiding the offset error of the RESET switch. While these techniques are shown with only one peak detector configuration, they can be applied to others.

Droop caused by an operational amplifier input bias current is effectively reduced by introducing a compensating source of droop developed with the other input current of the amplifier. This compensation is similar to that discussed with Fig. 3.22 and previously applied to sample-hold circuits.[5] An application of this technique to the peak detector of Fig. 3.20 is shown in Fig. 3.24. Note that a feedback capacitor which matches the storage capacitor has been added to A_2. If the input bias currents of A_2 are matched, they will produce equal but opposing voltage changes on the two capacitors for zero change at the output. The input bias currents seldom exactly match, but one or the other of the two capacitors can be experimentally increased to achieve very close compensation. Since both capacitors develop voltage drops, they create a common-mode voltage at the amplifier input. This common-mode voltage can overload the amplifier if not eventually removed by reset. Both capacitors of this circuit must be reset.

Leakage from the detector diode can be a greater source of droop than amplifier input current when FET operational amplifiers are used. In the HOLD mode the detector diode is reverse-biased to disconnect the input amplifier from the storage capacitor, and this reverse bias induces leakage current. Two methods for reducing this leakage are illustrated here. First, a JFET can be diode-connected to replace the detector diode as in Fig. 3.25. Generally, the JFET will provide far lower leakage than most economic diodes. Alternatively, a leakage decoupling circuit can be added such as R_2 and D_3 in Fig. 3.26. In the HOLD mode D_1 is still reverse-biased, but its leakage is conducted by R_2 rather than through the high reverse impedance of D_3. This leakage creates only a small voltage drop on R_2; and since R_2 is bootstrapped from the output, D_3 will be zero-

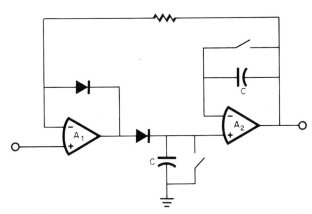

Fig. 3.24 To compensate a peak detector for droop caused by the input current of A_2, a feedback capacitor is added to produce a counteracting droop.

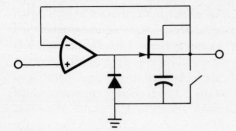

Fig. 3.25 Droop can be reduced through the use of a JFET as a low-leakage detector diode.

biased. Thus, very little leakage will flow through D_3 to drain the capacitor.

A similar decoupling technique removes the major leakage of the RESET switch to further reduce droop error. To provide this decoupling, Q_2 and R_2 are added to the reset circuit in Fig. 3.27. When the basic RESET switch Q_1 is off, it conducts a drain-source leakage current and a small drain-gate leakage. These leakages would drain the storage capacitor if the drain of Q_1 were connected directly to the capacitor. Switch Q_2 blocks these leakage currents and diverts them through R_2. Of course, the leakage of Q_2 now drains the capacitor, but this leakage is greatly reduced by bootstrap bias through R_2. With the switches off, R_2 holds the source of Q_2 very near the output voltage. Only a negligible voltage is developed on R_2 by the leakage of Q_1. Because the output voltage essentially equals the voltage at the drain of Q_2, this FET has zero drain-source bias. This zero bias removes the drain-source leakage of Q_2, leaving only the small gate-drain leakage as a drain on the capacitor. Addition of the leakage decoupling switch makes the net switch ON resistance twice as large and doubles the reset error described below.

The ON resistance of the RESET switch introduces a reset offset that may or may not cause error. Since the RESET switch shorts the output of

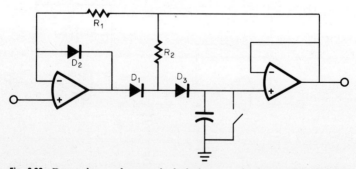

Fig. 3.26 Droop due to detector diode leakage can be removed through the use of bootstrapping feedback that holds detector diode bias at zero when the diode is not conducting.

an amplifier, it draws current from the amplifier. This current develops a voltage with the switch ON resistance and prevents complete discharge of the holding capacitor. However, the residual capacitor voltage does not necessarily constitute an error, because peak detector feedback limits the residual voltage.

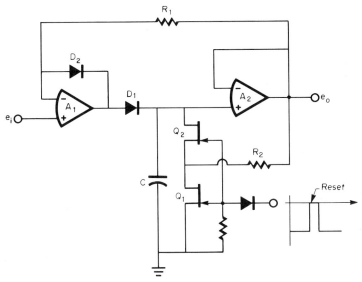

Fig. 3.27 Droop due to RESET switch leakage is reduced by decoupling the leakage away from C with switch Q_2 and bootstrap bias from R_2.

This can be seen by referring to the previous circuit of Fig. 3.27. If the reset voltage remaining on C were greater than the input signal, feedback would drive the output of A_1 negative, turning off D_1. With D_1 off, no current is supplied to the RESET switch by A_1, and the switch can discharge C to the input signal level or to zero, whichever is greater. A positive input signal that would permit a voltage to remain on C would induce the same or greater capacitor voltage when the reset ended. At the end of the reset the output would not be greater than the input signal; so no error would exist.

In practice the reset offset does sometimes create an error because of the time required to discharge the capacitor. The discharge cannot follow a rapid signal decrease that might occur just as the reset period ends. As a result, the output voltage after reset could be nonzero and greater than the input level. Where rapid signal changes could permit such an error, two other reset circuits can be used. In Fig. 3.28 a second switch Q_2 is used to shunt the amplifier output current before it reaches the RESET switch Q_1. None of the amplifier output current will flow in Q_1 so long as the voltage

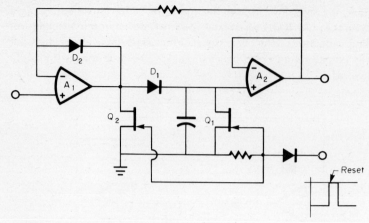

Fig. 3.28 Error from reset offset can be avoided by using a separate switch to shunt amplifier output current during reset.

developed on Q_2 is too low to forward-bias D_1. This condition is assured by choosing Q_2 for low ON resistance and for I_{DSS} greater than the maximum amplifier output current.

Alternatively, the reset error can be avoided by resetting the capacitor to a negative voltage as in Fig. 3.29. Here the source of the RESET switch is biased below ground with a resistor and zener diode. Now the RESET switch discharges the capacitor past zero to a negative voltage. This negative offset will later be removed by the first signal peak that is positive. A zener voltage is required that is greater than the FET pinchoff voltage but less than the negative supply voltage. Once again, the FET must have an I_{DSS} greater than the maximum amplifier output current.

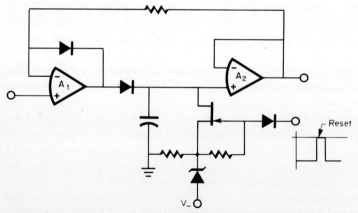

Fig. 3.29 By resetting a peak detector to a negative voltage rather than to zero, the error effect of RESET switch offset is avoided.

3.2.4 Improving peak detector speed

As mentioned before, the most serious response limitation of a peak detector is its acquisition time. This time required to switch from the HOLD mode to the SAMPLE mode and to acquire a new peak is dependent on several factors. These include the time required to switch between modes, the rate at which the storage capacitor can be charged, and the circuit settling time. In addition to basic settling time considerations, a settling response without overshoot is required of the input amplifier to avoid overcharging the storage capacitor.

The charging rate limit of the storage capacitor is controlled by a fundamental compromise between short acquisition time and long hold time. Ideally, a sample-hold circuit should require only a very short acquisition time and should be capable of a very long hold time. The hold time is limited by the output droop resulting from storage capacitor discharge by parasitic currents as described above. Large capacitances are desirable to reduce droop from parasitic discharge, but large capacitance increases the required sample time, resulting in a compromise between the two time characteristics. One way to avoid this compromise is to use two cascaded peak detector circuits. With this approach the first circuit has a small storage capacitance for rapid acquisition. The associated high droop rate is circumvented by the second peak detector, which acquires the stored voltage from the first circuit before significant droop occurs. For this transfer the second circuit can have a longer sample time and a larger capacitance for long hold time.

Obviously, another way to increase the capacitor charging rate and decrease acquisition time is to boost the capacitor charging current. To boost the charging current from the driving amplifier, a current booster can be added. This task is eased by the low duty cycle of the charging current in short sample times, which permits use of transistors of lower power capability. In peak detector circuits the current boosting can often be achieved by simply replacing a rectifying diode with an emitter follower. Such a modification is shown in Fig. 3.30 for a common peak detector circuit.[5] When the input signal to this circuit slightly exceeds the stored voltage established at the circuit output, the output of A_1 swings positive to turn on the transistor and further charge C. The charging current is primarily limited by the collector resistor instead of by the output current capability of A_1. Charging continues to make the output voltage follow the input signal until this signal begins to decrease. Since the transistor cannot then supply the opposite polarity current for discharge, it becomes reverse-biased, disconnecting the storage capacitor from the signal. As a result, the capacitor voltage remains at the level of the most recent signal peak unless reset. If the reverse-bias voltage on the transistor is capable of driving the emitter-base junction into breakdown, a protection diode must be added in series with the transistor base. With this added diode,

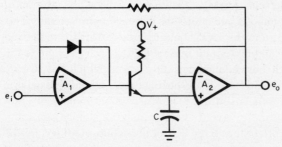

Fig. 3.30 A peak detector with an emitter follower in place of a rectifying diode has a greater capacitor charging rate limit.

the leakage decoupling technique of Fig. 3.26 is available solely for the addition of a resistor (shown there as R_2) to the base of the transistor.

The time required for a peak detector to switch from the HOLD mode to the SAMPLE mode is determined by the slewing rate and gain-bandwidth product of the input amplifier. For small signals a serious switching time limitation is imposed by the circuit diodes. Switching requirements can be seen from the previous circuit of Fig. 3.30. To acquire a new peak level, the output of A_1 must swing a voltage equal to the input transition, and then a voltage equal to two diode drops $2V_f$ to turn off the diode and turn on the transistor. For large input signals this output swing is slew-rate-limited, and the slewing time is somewhat increased by the added $2V_f$ voltage swing.

The time required for this added swing represents an increasingly greater portion of the total switching time for lower signal levels. Where the input signal transition is small compared with $2V_f$, the switching time is essentially that required for the amplifier output to traverse $2V_f$. Small signals will not drive the amplifier to its slewing-rate limit, and the rate of change of the amplifier output voltage will equal the rate of change of the input signal multiplied by the open-loop gain of the amplifier at the effective signal frequency $A(f_i)$. The transition time will then be the time required for the input signal to traverse a voltage of $2V_f/A(f_i)$. Since this gain decreases with increasing frequency, faster signals must have larger amplitude to develop the required $2V_f$ output swing. That portion of the input signal required to develop this junction switching output swing represents gain error, and it greatly limits the detection of small, fast peaks.

Significant reduction in switching time can be achieved by bootstrap-clamping the input amplifier, by removing phase compensation during the switching transition, and by selectively adding gain during the switching transition. For large signal applications, bootstrap clamping of the input amplifier reduces the amplifier voltage swing requirement to decrease switching time. This is provided by D_3 and R_1 in Fig. 3.31. As the input signal decreases from a peak, D_3 clamps the input of A_1 at one diode drop below the stored output level. Otherwise, A_1 would follow the input signal down in voltage and then have to slew back up with the next peak. But with the clamp the output of A_1 is only required to swing $2V_f$ plus the difference between the new peak level and the stored voltage. This can greatly reduce switching time and permit detection of significantly shorter duration peaks.

However, the peaks must be repetitive or recurrent in order to benefit from the bootstrapped clamp. A single pulse has no preceding peaks to set up a bootstrap bias, but a repetitive or recurrent signal will gradually build up the bias until the last increment of the peak is acquired. Also provided by the bootstrap bias are some accuracy improvements. The reverse bias on D_1 is now limited to V_f, giving reduced diode leakage and droop. Reduced voltage on R_2 permits use of a much smaller resistor that helps avoid overshoot. Overshoot is also reduced by the decreased amplifier swing.

Peak detector mode switching can also be made faster by removing the response-limiting amplifier phase compensation during the switching transition. During that portion of the switching transition for which neither diode is on, the feedback loop around the input amplifier is open. In this open-loop state most operational amplifiers require no phase compensation and would have a much higher slewing rate and high-frequency gain without the compensation. Then, by switching out the phase compensation for the open-loop state, frequency response limitations are significantly reduced.

For operational amplifiers with external, Miller effect, or feedback phase

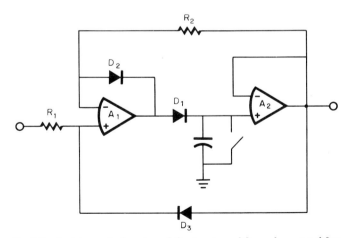

Fig. 3.31 Bootstrapped clamping of the input amplifier reduces amplifier voltage swing requirements to decrease peak detector switching time.

compensation,[1] the described operation is achieved by the circuit of Fig. 3.32. With this modified configuration the output of A_1 is connected to the compensation capacitor C_2 only when the switching transition is completed and D_1 is on. As a result, phase compensation is connected when needed for stability but is not connected when it would unnecessarily limit the switching transition time. Note that D_2 is connected to ground rather than in feedback around A_1. This removes feedback from A_1 for the entire signal rise up to the turn-on point of D_1 and removes the need for phase compensation during this transition. Since the grounded diode clamp shorts the output of A_1, that amplifier should be chosen for very rapid overload recovery following output shorting.

Alternatively, the rate of change of the critical amplifier output signal can be boosted by amplifying that signal. However, addition of gain in feedback circuits is accompanied by the need for added, stabilizing phase compensation, and the resulting gain-bandwidth product and slew rate are not increased. Fortunately, the feedback loop is open during the switching transition; so the added phase compensation is not needed when the added gain is desired. If gain can be selectively added only during the open-loop, switching transition, then switching speed will be boosted, and feedback stability following switching will be maintained.

The speed boosting operation described above is achieved with the circuit of Fig. 3.33. With the stage formed by Q_1 through Q_4, gain is added during the switching transition. During this transition, both diodes are off and do not shunt the output of the added stage. Following the transition, one diode conducts, heavily shunting the high output impedance of the stage and dropping its gain. Thus, the high output impedance of the added stage ensures that gain is added only during the switching transition.

To further boost the response, the added stage is driven from the power-supply current drains of the operational amplifier. With this circuit

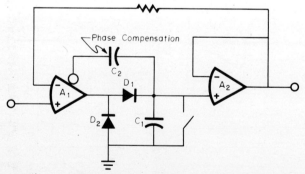

Fig. 3.32 Improved peak detector response is achieved with feedback phase compensation that is connected only when mode transition is complete and the detector diode is forward-biased.

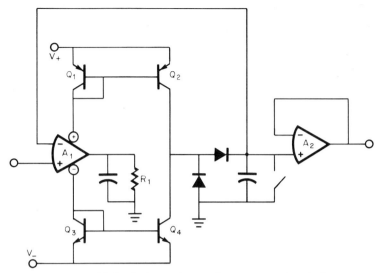

Fig. 3.33 Gain is added only during the switching transition by a high output impedance stage to reduce peak detector response time.

the output swing required of the amplifier is greatly reduced. This lower output swing does not reach the amplifier slew-rate limitation until a much higher frequency than the amplifier full-power response. A low resistance load on the amplifier draws rated output current for only a small output swing. Current drawn from the amplifier output must be supplied through Q_1 or Q_3. This produces a matching current in Q_2 or Q_4, which then drives the switching feedback impedance. Accurate matching of like transistors can be achieved with monolithic pairs. Alternatively, unmatched transistors can be used if emitter degeneration resistors are added to stabilize biases.

Note that the feedback to A_1 is taken from the storage capacitor rather than from the output of A_2. This sacrifices the error reduction achieved with A_2 inside the loop and requires that A_1 have low input bias current and high input impedance under overload. However, this feedback connection permits greater utilization of the new response capabilities of A_1. Unless limited by the capacitor charging rate, increased slewing rate is provided by the added stage in the SAMPLE mode as well as during the switching transition. If this slewing rate is greater than that of A_2, A_2 introduces delay in any feedback supplied to A_1. This delay would cause A_1 to overshoot and overcharge the storage capacitor.

3.3 Voltage Discriminators

To choose one of several signals on the basis of magnitude, voltage discriminators can be used to select the maximum, median, or minimum

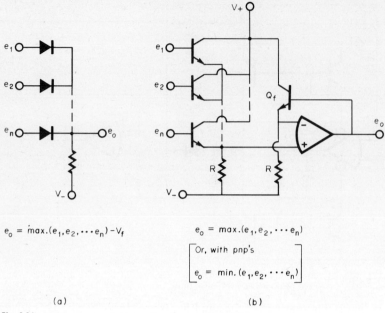

Fig. 3.34 Maximum or minimum signals are selected for connection to a circuit output by means of gating diodes or transistors.

signal. Such selection is performed by circuitry that first compares the signal magnitudes and then connects the appropriate signal to the circuit output. In simplest form, a maximum or minimum discriminator consists of a gating diode for each signal and a bias resistor as shown in Fig. 3.34a. For the diode orientation shown, the maximum input signal will control the output voltage. That signal will forward-bias its diode path to the output and reverse-bias the other diodes. If the diodes and the bias voltage polarity are reversed, the minimum, or most negative, signal will control the output voltage in an analogous manner.

With this elementary circuit large errors are introduced that can be removed by operational amplifier feedback. The major error consists of the signal loss to forward-bias the conducting diode. In addition, the circuit input resistance is not high, nor is the output resistance low; so loading errors can be significant. Each of these deficiencies is removed with the transistors and amplifier of Fig. 3.34b. High input resistance is developed by the buffering of the input transistor current gains, and low output resistance is provided by the amplifier output. Signal gating is performed by the emitter-base junctions of the transistors instead of diodes, and the emitter-base breakdown voltages must be large enough to support input voltage differences. The signal lost to forward-bias an emitter-base junction is accurately compensated for by the emitter-base voltage of Q_f. The latter transistor increases the output signal by a voltage that tracks the signal loss on the input transistor. Such compensation results from tracking current levels in the input transistors and Q_f. The two currents are determined by voltages on the two resistors, which are in turn controlled by bias and signal. Because of operational amplifier feedback, the two resistor voltages will always be equal, to maintain the desired transistor current match. If the input transistors are matched to Q_f over the range of currents conducted, accurate compensation for input signal loss will be developed. Also required is a close matching of the two resistors to result in equal signal currents.

Alternatively, the circuitry of Fig. 3.34b can be used for selection of minimum signals rather than maximum signals. This is achieved by substituting pnp transistors for the npn's and by reversing the bias potential polarities.

Higher accuracy maximum or minimum selection results from enclosure of the gating diodes in separate amplifier feedback loops. In this way the voltage drops of the diodes are not supported by the signal, but rather by

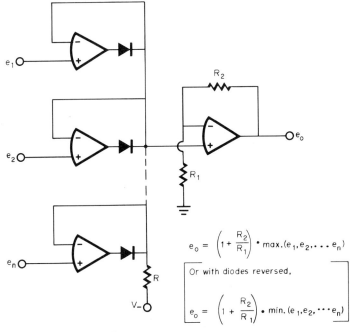

$$e_o = \left(1 + \frac{R_2}{R_1}\right) \cdot \max.(e_1, e_2, \ldots e_n)$$

Or with diodes reversed,

$$e_o = \left(1 + \frac{R_2}{R_1}\right) \cdot \min.(e_1, e_2, \ldots e_n)$$

Fig. 3.35 Diode voltage errors are removed from maximum or minimum selector circuits by enclosing the diodes in amplifier feedback loops.

the amplifier output; so the diode voltage errors are reduced by the amplifier gains. This error reduction, plus the availability of circuit gain, is provided by the circuit of Fig. 3.35. Each input amplifier attempts to control the voltage at their common output, but only one can succeed since the diodes block one polarity of their output currents. Only that amplifier attempting to establish the largest output voltage maintains control over its output current and, therefore, of the output voltage. The controlling amplifier is the one with the maximum input voltage, and its output raises the inverting inputs of all other amplifiers to a higher level than required for their feedbacks. As a result, the other amplifier outputs swing to their negative saturation levels.

Only one polarity of output current can be supplied by the input amplifiers; therefore an additional buffer amplifier is added, as shown, if the opposite polarity output current is required. For grounded loads, the input amplifiers alone can only supply current to the loads in response to positive input voltages. If the maximum input signal could be negative, little current could be supplied to a grounded load without the additional output buffer shown. Also required because of the diode blocking is the resistor R, which conducts the input currents of the amplifiers if those currents are of a polarity not conducted through the diodes. Further, the

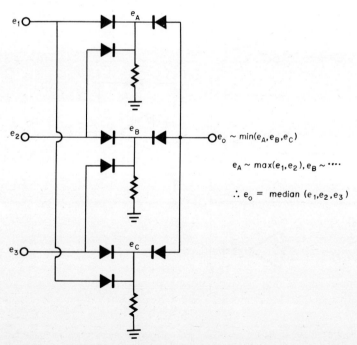

$e_o \sim \min(e_A, e_B, e_C)$

$e_A \sim \max(e_1, e_2), e_B \sim \cdots$

$\therefore e_o = \text{median}(e_1, e_2, e_3)$

Fig. 3.36 To select a median signal, maximum selections are performed followed by a minimum selection.

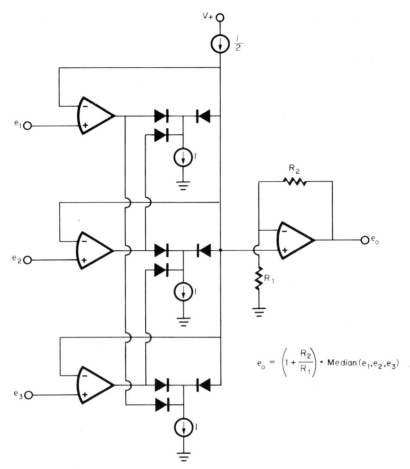

Fig. 3.37 To reduce error from diode voltage drops, the diodes of the previous median selector circuit are enclosed in amplifier feedback loops.

diode switching introduces a bandwidth limitation like that described in Chapter 5 for absolute-value circuits.

Another voltage discriminator function is that of selecting a median-level signal rather than a maximum or minimum. For three signals, median selection is performed by the circuit of Fig. 3.36. Each signal is first compared against the other two by maximum selectors. Two of the resulting signals e_A, e_B, and e_C will be dominated by the maximum of the three input signals. The remaining maximum selector output will be controlled by the median input signal. By means of the minimum selector, this median signal can be discriminated from the outputs of the maximum selector circuits.[10]

This operation is performed by the circuit of Fig. 3.37 using operational amplifier feedback to remove signal loss from diode bias. Maximum selection is again performed by the diodes having their cathodes oriented to the right, and minimum selection is performed by the oppositely oriented diodes. Each input amplifier attempts to establish its input voltage at the common output terminal, but only one can do so. That will be the amplifier which produces the lowest cathode bias to the right-hand diodes. Lowest bias results from the smallest of the maximum selections performed by the left-hand diodes. Thus, output control is maintained by the median-level signal.

While the resulting output signal voltage e_o accurately tracks the voltage selected because of the amplifier feedback, the feedback does not remove selection errors. The maximum and minimum comparisons are performed within the amplifier feedback loops and are not adjusted by the feedback. Errors in these comparisons are introduced by any diode or current-source mismatch. Either mismatch will alter the results of the maximum and minimum comparisons such that the wrong signal may be selected. Also developed by the diodes is a bandwidth limitation like that described for absolute-value circuits in Chapter 5.

REFERENCES

1. G. Tobey, J. Graeme, and L. Huelsman, *Operational Amplifiers; Design and Applications*, McGraw-Hill Book Company, New York, 1971.
2. M. Strange, Variable Threshold Circuit Separates Sync Pulses from Composite Video Signal, *Electron. Des.*, October, 1972.
3. J. Graeme, Comparator Has Precise, Voltage–Controlled Hysteresis, *EDN*, August 20, 1975.
4. J. Graeme, Voltage Comparator Circuit Gives Audio Alarm When Tripped, *Electron. Des.*, May 10, 1975.
5. J. Graeme, *Applications of Operational Amplifiers: Third-Generation Techniques*, McGraw-Hill Book Company, New York, 1973.
6. J. Graeme, Window Comparator Needs Only One Op Amp, *Electronics*, September 5, 1974.
7. J. Graeme, Tame Comparator Hysteresis to Make All Those Promises Come True, *EDN*, May 5, 1975.
8. J. Graeme, Peak Detector Advances Increase Measurement Accuracy, Bandwidth, *EDN*, September 5, 1974.
9. J. Graeme, Getting Inside a Peak Detector to Make It Do the Job, *Electronics*, November 14, 1974.
10. A. Moses, Midvalue Selector Doubles as a Precise Voltage Limiter, *Electron. Eng.*, December, 1970.

4
SIGNAL CONDITIONERS

The high precision and versatility of operational amplifiers are well applied in controlling signal characteristics such as magnitude, frequency content, or even the signal frequency. Among these applications are voltage regulation, filtering, and frequency multiplication as provided by the circuits described in this chapter. Voltage regulators are presented that simplify common linear regulation or switching regulation and that provide more specialized regulator operation such as digital control of reference voltages. Further regulator utility is achieved by using operational amplifiers with regulators to expand output capability. Active filters lend simplicity to filter design, particularly with the modified state-variable configurations presented. Also made available with active filters is voltage-controlled response for automatic testing, and this capability is extended to digital control. Frequency multiplication makes possible the derivation of higher frequency timing signals from a given signal, and operational amplifier circuit approaches to this task are illustrated.

4.1 Voltage Regulators

One of the most common signal conditioners, the voltage regulator, transforms a varying signal into a constant voltage either for power-supply or reference voltage applications. The required function is well served by

100 Designing with Operational Amplifiers

the high gain and power-supply rejection of operational amplifiers, because these characteristics ensure precise voltage regulation and high line rejection. In addition, operational amplifiers can be used for more specialized regulator functions or to extend the range of performance of fixed regulators.

4.1.1 General-purpose circuits Numerous operational amplifier configurations provide voltage regulation.[1,2] Additional circuits presented here include simple voltage references, a high-precision regulator with variable output, and ripple cancellation techniques. The simple voltage reference is intended for improvement over voltage references established by voltage dividers from power-supply voltages. Such divider-derived references are sensitive to power-supply variations and loading and have poor power efficiency. Each of these deficiencies is removed with an operational amplifier reference that develops an output voltage in response to adjustment of the amplifier voltage offset null as in Fig. 4.1. With this approach, a reference voltage can be established at any level within the output voltage range of the amplifier, and that reference voltage is insensitive to power-supply variations or output loading.

However, this reference voltage will generally be temperature-sensitive because of the effect of the offset nulling upon the amplifier input offset voltage drift. Typically, this drift is changed about 3 μV/°C for each millivolt of input offset voltage produced by the null control.[2] This represents a temperature coefficient of 0.3%/°C for the reference voltage established. Even greater temperature coefficients result if too great an offset is generated with the null control because of amplifier drift sensitivites made more significant by altered internal amplifier biases. For this reason, it is desirable to limit the input offset voltage change to approximately 30 mV, which can then be boosted to the desired reference voltage level by the circuit gain.

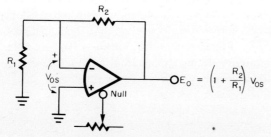

Fig. 4.1 For simple voltage reference requirements the input offset voltage of an operational amplifier can be adjusted to develop the desired voltage level using high closed-loop gain.

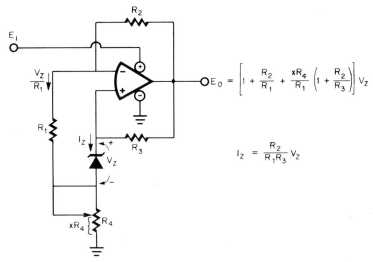

Fig. 4.2 Precise output voltage control for this variable regulator is aided by a zener bias that is held constant for minimum zener voltage drift even though the output voltage may be varied.

For more stable reference voltages or for general regulator applications, other operational amplifier configurations are available. Many of these operate from the single-polarity unregulated power supply required for typical power-supply regulators. One of the more precise of these operational amplifier circuits biases a reference zener diode from the stable regulator output for improved ripple rejection.[1] More traditionally, a regulator reference zener is biased from the unregulated input power, where ripple and other voltage variations alter the current supply to the zener and thereby introduce variations in the regulator output voltage. While reference bias from the regulator output removes this source of error, it subjects that bias to changes introduced in varying the regulator output voltage. Since a variation in zener diode bias current changes the zener voltage temperature coefficient, adjustment of regulator output voltage would alter regulator drift.

That effect, too, can be avoided[3] with a modification illustrated in Fig. 4.2. Variable control over output voltage is provided by R_4 without altering the zener diode current I_Z. Fixing I_Z is the voltage established on R_3, which must equal that on R_2 to maintain zero voltage between the amplifier inputs. This zero-voltage constraint also forces the voltage on R_1 to match the zener voltage, and the resulting current in R_1 produces a constant voltage on R_2 to be matched by the voltage on R_3. As a result, the zener diode current is determined by the zener voltage and three resistors independent of the output voltage.

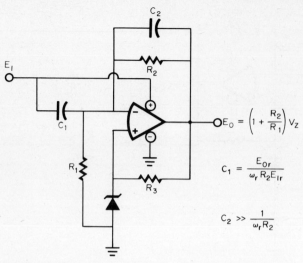

Fig. 4.3 Regulator output ripple can be reduced by supplying a compensation signal to the amplifier from the unregulated input voltage.

Because the currents in the zener diode and resistor R_1 are constant, they develop a constant voltage on the resistance set by R_4. By varying that resistor, the associated voltage can be adjusted to control the output voltage as expressed in the figure. While the zener alone would provide a constant current to R_4, it is necessary to also connect R_1 to this control potentiometer for feedback stability. Instability would result if the positive feedback factor of the zener bias connection were to exceed the negative feedback factor of R_1 and R_2. With the connection shown, R_4 does not alter the dominance of the negative feedback.

As mentioned, the output ripple is greatly reduced by deriving the reference diode bias from the stable regulator output voltage. Remaining, however, is the smaller ripple produced by the power-supply rejection error of the operational amplifier. Even this ripple can be largely removed by adding a ripple cancellation signal to the amplifier[4] as in Fig. 4.3. The ripple cancellation signal is supplied from the input ripple voltage to the amplifier by C_1 to develop a counteracting output ripple. Cancellation occurs if the output ripple is originally in phase with the input ripple, as generally occurs. By appropriate selection of C_1, the net output ripple can be readily reduced by a factor of 4, or with more careful selection by a factor of 10. For accurate ripple cancellation, C_1 should be

$$C_1 = \frac{E_{Or}}{\omega_r R_2 E_{Ir}}$$

where E_{Or}, E_{Ir}, and ω_r are the ripple output voltage, input voltage, and frequency, respectively. Also coupled to the amplifier by C_1 are spurious transients in the unregulated input voltage; so C_2 is added to lower the amplifier gain at higher frequencies.

Ripple rejection can also be improved by biasing the regulator amplifier, along with the reference, from the regulator output. This removes the power-supply rejection error of the amplifier from the regulator output. In order to use the regulator output voltage as a power supply for the operational amplifier, the regulator output voltage must be greater than that required from the operational amplifier output. For this reason, voltage-level-shifting circuitry is required between the amplifier output and that of the regulator as provided by Q_3 in Fig. 4.4. Also, an output-buffering pass transistor is added, shown as Q_1, with a current-limit transistor Q_2.

Output voltage is sensed by the amplifier through the voltage divider formed with R_1 and R_2, where it is compared against the reference voltage V_{Z2}. If the output voltage were zero, the reference diode would have no bias; so it would present a zero voltage reference. However, zero output is not a stable state because Q_1 is then biased on by R_3 to raise the output voltage. The same zero output zero-reference condition might appear to exist for the previous two circuits that had references biased from the output. But in those cases the amplifier was biased from the input voltage. When so biased, the amplifier always turns on, and it prevents zero voltage from being a stable output state by virtue of the fact that the amplifier output cannot swing all the way to zero volts.

It is this same fact that requires addition of D_{Z1} to shift the output of the amplifier in Fig. 4.4 away from ground. A zener diode is chosen instead of

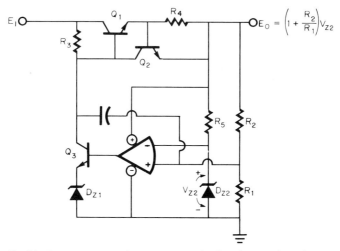

Fig. 4.4 Even greater ripple rejection results from biasing the reference amplifier from the regulator output along with the reference diode.

a resistor for this purpose in the interest of higher ripple rejection. If a resistor were used, it would have a large ripple voltage swing in order to supply the ripple voltage across R_3. The result would be a large ripple voltage on the amplifier output, which in turn would permit a ripple on the regulator output associated with the gain error signal at the amplifier input.

When the reference diode is biased from a regulator output for ripple rejection as in the previous circuits, the output voltage must remain greater than that of the reference diode. This prevents use of the technique at lower voltage levels for which zener diodes are not available or not precise. Instead, a low-voltage reference can be formed with transistors by making use of the highly predictable law of the junction[1] governing the current-voltage characteristics of semiconductor junctions. One such low-voltage reference is used with the regulator of Fig. 4.5. If one merely selects the appropriate current ratio for Q_1 and Q_2, a stable, repeatable current is produced in Q_1 for use in setting the reference voltage on R_3. The resulting voltage need only be large enough to bias the noninverting input within the common-mode range of the amplifier. This bias requirement results from single-supply operation of the operational amplifier where one of its supply terminals is returned to ground. For linear operation, the amplifier input voltage must be above that of the negative supply terminal, but only by 2 or 3 V for many operational amplifiers. This permits regulator output voltages as low as 3 V.

To ensure a stable reference voltage, the current supplied by Q_1 is controlled by Q_2 and Q_3 in a manner that limits total tolerance to a few percent and the temperature coefficient to about $0.05\%/°C$. Initial current control

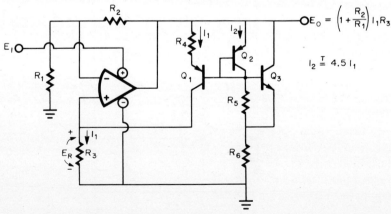

Fig. 4.5 For output voltage levels below those of zener diodes, a voltage regulator can make use of the highly predictable characteristics of transistor junctions to establish a low-voltage reference that can be biased from the regulator output.

is provided by Q_3, which serves as a predictable but temperature-sensitive current source to Q_2. Essentially, the current delivered to Q_2 is that conducted through R_5 as controlled by Q_3, and that current I_2 equals V_{BE3}/R_5. All other current from R_6 is diverted through Q_3. Because I_2 varies with temperature in response to the variation in V_{BE3}, I_2 cannot be used direct to set the reference voltage. However, a compensating thermal drift can be added to hold I_1 constant by merely making the magnitude of I_1 a certain fraction of I_2. This fraction is controlled by R_4 and is typically 1:4.5 for silicon transistors at room temperature.

For a given application this typical current ratio can be used to design the circuit, or more accurate design can be performed starting with the junction equation. From that equation, the emitter-base voltage difference created by unequal currents in Q_1 and Q_2 is[1]

$$\Delta V_{BE1,2} = \frac{KT}{q} \ln \frac{I_2}{I_1}$$

Since this emitter-base voltage difference is the voltage impressed on R_4 to establish I_1, it is necessary to make it independent of temperature:

$$I_1 R_4 = \frac{KT_1}{q} \ln \frac{I_2(T_1)}{I_1} = \frac{KT_2}{q} \ln \frac{I_2(T_2)}{I_1}$$

Temperature independence is approximated by using the decrease in I_2 with temperature to compensate the normal positive drift of ΔV_{BE}. To find the appropriate level of I_2, the -2 mV/°C variation of V_{BE3} is used to define

$$I_2(T_2) = I_2(T_1) - \frac{(2 \text{ mV/°C})(T_2 - T_1)}{R_5} \qquad T_2 > T_1$$

Using the last two equations, a room-temperature level for I_2, and then a resistance value for R_4, can be derived for a given level of I_1.

4.1.2 Switching regulators Switching regulators offer greatly improved power efficiency and small size at the expense of some increase in regulator complexity and in electromagnetic interference. High power efficiency results from operating the regulator pass transistor as a switch so that it conducts current primarily when its voltage drop is very low. This reduces power dissipation in the transistor for reduced heat sinkage as well as increased efficiency. Reduced heat sinking makes for a more compact regulator as does the opportunity for reducing the size of filtering elements. Smaller ripple filtering components can be used with a switching regulator when that regulator is operated at a switching frequency well above the line frequency.

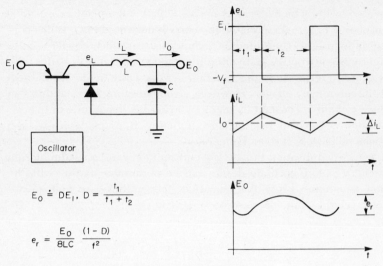

Fig. 4.6 A switching regulator connects an input voltage to an LC filter with a controlled duty cycle to develop a lower level output voltage.

Basically a switching regulator consists of a switch that connects an input voltage to an LC filter in response to an oscillator drive as in Fig. 4.6. When the transistor switch is on, it connects the input voltage to the filter; and when it is off, the diode provides a current return for the inductor. As a result of the switching, the voltage presented to the filter is a rectangular waveform with a duty cycle D controlled by the oscillator. The filter produces an output voltage that is ideally a dc voltage equal to the average value of the rectangular waveform, or

$$E_O \doteq DE_I \qquad E_I \gg V_f$$

where D is the duty cycle

$$D = \frac{t_1}{t_1 + t_2}$$

Thus, by an appropriate choice of the duty cycle, the output voltage can be set to a wide range of levels for a given input voltage.

In the design of a switching regulator, the duty cycle, oscillator frequency, inductance, and capacitance are selected considering output voltage, cost, efficiency, and output ripple. The duty cycle is fixed by the ratio of the input to output voltages as expressed above. Lower cost results with smaller values for L and C, but this raises the frequency that must be used. At higher frequency the transistor switching time represents a greater portion of the switching period, and the switching transition represents the time of maximum power dissipation for the transistor. So efficiency is degraded at high frequencies. For a given efficiency requirement, a maximum operating frequency can be defined and then examined against the noise tolerance of the intended application. Both electromagnetic and audible noise can be generated by switching regulators. It is desirable to set the operating frequency above the audible range and, if possible, above the frequency of signals occurring in the circuits to be supplied by the regulator, for ease of noise filtering.

Once the operating frequency is defined, the values for L and C are determined considering the output ripple that can be tolerated. Ripple is related to that portion of the load current that must be supplied by the capacitor. That current is the difference between the inductor current i_L and the output current I_0. As shown in Fig. 4.6, i_L rises and falls about the level I_0 as the input voltage is alternately connected to or disconnected from the filter. Associated with this difference current is the change in capacitor charge that equals the area between the i_L curve and the I_0 level.[5] This area is composed of right triangles with height $\Delta i_L/2$ and with lengths of $t_1/2$ and $t_2/2$. Total area between the two curves for one cycle is

$$\text{Area} = \Delta Q = \frac{1}{2} \frac{\Delta i_L}{2} \left(\frac{t_1}{2} + \frac{t_2}{2} \right) = \frac{\Delta i_L}{8}(t_1 + t_2)$$

This change in charge equals the change in capacitor voltage, or ripple e_r, times the capacitance; so

$$e_r = \frac{\Delta Q}{C} = \frac{\Delta i_L}{8C}(t_1 + t_2) = \frac{\Delta i_L}{8Cf}$$

Ripple is then defined in terms of the change in inductor current Δi_L, which is related to the inductance and the change in inductor voltage. Essentially all of the voltage e_L appears across the inductor, assuming that the ripple is small in comparison; so

$$L \frac{\Delta i_L}{t_2} \doteq E_0 \qquad \Delta i_L \doteq \frac{E_0 t_2}{L} \qquad \text{for } E_0 \gg V_f$$

Combining this with the previous equation defines the requirement for L and C in terms of the fractional ripple, the switching frequency, and t_2:

$$LC = \frac{E_0}{8e_r} \frac{t_2}{f} = \frac{E_0}{8e_r} \frac{1-D}{f^2}$$

The elementary switching regulator of Fig. 4.6 has large output errors that can be removed with feedback control of the switching. Large errors result if the input voltage and the duty cycle are not well controlled. To control these characteristics, the input voltage must be regulated, the oscillator must be stabilized, and the switching frequency must be low

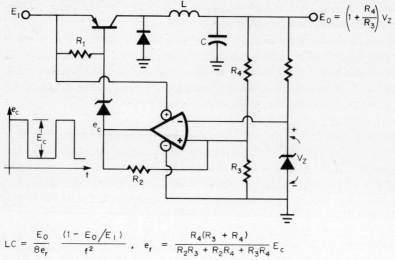

$$LC = \frac{E_O}{8e_r} \frac{(1 - E_O/E_1)}{f^2}, \quad e_r = \frac{R_4(R_3 + R_4)}{R_2 R_3 + R_2 R_4 + R_3 R_4} E_c$$

Fig. 4.7 Feedback control of a switching regulator output is provided by a comparator in this self-oscillating configuration.

enough that switching time variations are not significant. Rather than control each of these characteristics, it is simpler to use feedback control that adjusts the duty cycle to maintain a constant output voltage.

In the simplest case, feedback control is provided by a self-oscillating switching regulator like that of Fig. 4.7. This circuit uses an operational amplifier as a comparator to sense any output variation from a level related to a reference zener voltage. Whenever the output voltage falls below the design level, the comparator output switches to its negative state to turn on the switching transistor. This connects the input voltage to the filter in order to restore the regulator output voltage, and then the comparator drives the switch off again. To ensure adequate switch turnoff drive voltage, the comparator output voltage is level-shifted through a zener diode. Switching continues as the output ripple drives the comparator between its two output states.

In order to control the frequency of oscillation of a self-oscillating switching regulator, the comparator sensitivity to ripple is adjusted with hysteresis as supplied by the positive feedback of R_2. The amount of hysteresis is related to the peak-to-peak comparator output swing defined as E_c and will be

$$E_H = \frac{R_3 \parallel R_4}{R_2 + (R_3 \parallel R_4)} E_c$$

From this hysteresis an output ripple results of

$$e_r = \left(1 + \frac{R_4}{R_3}\right) E_H$$

While low ripple is desirable, a minimum level must be accepted with a self-oscillating switching regulator in order to achieve fast switching. Comparator switching speed is largely determined by the amount of positive feedback supplied; so hysteresis and therefore ripple are required to permit efficient switching. Once the level of output ripple is chosen, L and C are chosen to set the oscillation frequency using the previously derived equations

$$LC = \frac{E_0}{8e_r} \frac{1-D}{f^2} \qquad D = \frac{E_0}{E_I}$$

As before, the oscillation frequency is chosen from consideration of efficiency and noise, and duty cycle is defined by the ratio of input to output voltages.

To avoid the excess ripple inherent in a self-oscillating switching regulator, switching can be controlled by pulse width modulation as in Fig. 4.8. With pulse width modulation the duty cycle is controlled by the average of the output voltage rather than by the ripple about that average. The average voltage is sensed by the integrator error amplifier formed with A_2. If the average voltage at the junction of R_4 and R_5 differs

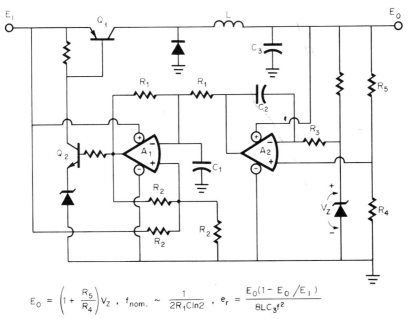

$$E_0 = \left(1 + \frac{R_5}{R_4}\right) V_Z, \quad f_{nom.} \sim \frac{1}{2R_1 C \ln 2}, \quad e_r = \frac{E_0(1 - E_0/E_I)}{8LC_3 f^2}$$

Fig. 4.8 Pulse width modulated control of a switching regulator avoids the higher ripple of the self-oscillating switching regulator.

from V_Z, the integrator output voltage will be changed in a direction to correct the duty cycle of the pulse width modulator formed with A_1. At equilibrium the integrator output holds at that level which maintains the appropriate duty cycle.

The simple pulse width modulator formed with A_1 is merely a basic square-wave generator modified for a modulation input[2] and for operation on a single power supply. The single power-supply operation is made possible by the bias provided from E_1 to the noninverting input of A_1. Modulation results from the current supplied to C_1 through a resistor from the output of the integrator. Because of this added current, the capacitor charging is boosted in one direction and decreased in the other. This reduces the time spent in one oscillator state and increases that spent in the opposite state; so the output waveform will be altered from its normal 50 percent duty cycle. At circuit equilibrium the duty cycle reaches that level appropriate for the specific input-output voltage ratio.

In selecting components for a specific application of the circuit of Fig. 4.8, the characteristics considered include those described with the elementary switching regulator plus other requirements of the integrator and pulse width modulator. As before, operating frequency is selected considering efficiency, noise, and filter element cost. The nominal frequency of the pulse width modulator output is

$$f_{nom} \approx \frac{1}{2R_1 C \ln 2}$$

However, this frequency is lowered by the modulation as much as 50 percent for either an increasing or decreasing duty cycle. Filter element values are chosen to limit the output ripple by means of the expression

$$LC_3 = \frac{E_O}{8e_r} \frac{1 - E_O/E_I}{f^2}$$

Because of the phase shift of the filter elements, there is a delay in output response following a feedback change in duty cycle. To avoid feedback-loop oscillation from this delay, the integrator time constant $R_3 C_2$ is selected for appropriate loop damping.

Components are chosen for the pulse width modulator with switching speed and bias considerations. A high-speed operational amplifier without its response-limiting phase compensation or a fast comparator should be used for A_1 to maximize switching speed. In addition, A_1 must maintain high input impedance under the input overload that is a normal state for this oscillator.[2] Switching speed is also boosted by the addition of a second switching transistor Q_2. That transistor has a zener diode emitter bias which level shifts it to the output voltage range of A_1.

4.1.3 Specialized voltage regulators

Other regulator functions finding fairly frequent applications are those of foldback current limiting and digitally programmable output voltage. Such digital control makes possible automatic sequencing of voltage sources in electrical testing. Foldback current limiting offers significant reduction in overload power dissipation for pass transistors. Under output overload, regulator current limiting results in decreasing output voltage. This voltage decrease is accompanied by an equal increase in voltage on a regulator pass transistor with the potential for increased power dissipation in these devices. With conventional constant-level current limits, the maximum dissipation in these transistors occurs when the output voltage is reduced to zero by a short circuit. Power-supply performance is greatly limited by the need to meet this condition. If output transistors and heat sinking are selected for this short-circuit case considering a constant current limit, they will typically handle only one-fourth their power capability under normal operating conditions.

By means of a foldback current limit the pass transistor dissipation under output overload can be held close to that of normal operating conditions. This permits output current supply of about four times that permissible with a constant-level current limit. As the name implies, foldback limiting produces a regulator output current-voltage response like that of Fig. 4.9. Note that the output current is reduced to lower levels as the output voltage decreases in the current-limit mode.

While this does greatly reduce power dissipation in the regulator, it also makes possible a latching mode that prevents turn-on to full output voltage. Latching can occur when the regulator supplies a current-source load such as that represented by the dashed load line. Load lines like that

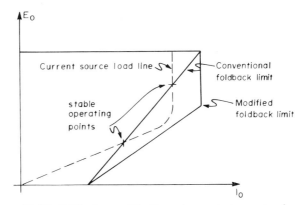

Fig. 4.9 Foldback current limiting reduces output current under overload to avoid increased regulator power dissipation, but latching in the foldback state is possible unless the foldback is modified.

shown are commonly produced by the quiescent current turn-on characteristics of many electronic circuits. Since this load line intersects the conventional foldback response, it establishes stable operating points below the full regulator output voltage. Where this can occur, a modified foldback limit[6] can be used to avoid latching while still achieving much of the reduction in pass transistor dissipation under overload. As shown in Fig. 4.9 this modified foldback limit can be tailored to avoid intercept with many practical current-source load lines.

An implementation of this modified current limit is shown in Fig. 4.10. Here a zener diode–referenced clamp adds a current-limit-altering signal for output voltages below a certain level. Conventional constant current limiting[2] is often provided with combinations like Q_2 and R_3. When output current develops a voltage on R_3 equal to the emitter-base voltage of Q_2, that transistor turns on to limit the base drive current to Q_1. This limit is modified under output overload conditions by D_{z1}, D_1, R_1, and R_2. Under overload, Q_2 initially provides constant current limit, but the resulting drop in output voltage eventually forward-biases D_1 so that the current limit is reduced. Once D_1 begins conducting, further reduction in output voltage results in current flow through R_2 from R_1. This current develops a voltage on R_2 that helps supply the emitter-base bias of Q_2; so less voltage will be allowed on R_3 in the limit state. Thus, output current is reduced from its initial limit level, and further reduction will result as the output voltage continues to decrease.

The second specialized regulator considered is digitally programmed

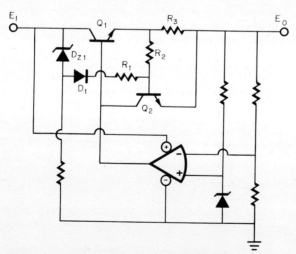

Fig. 4.10 A modified foldback current limit like that of Fig. 4.9 is provided for a regulator through the action of D_{z1}, D_1, R_1, and R_2 together with a conventional current-limiting transistor.

Signal Conditioners   113

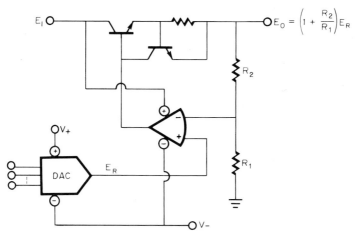

Fig. 4.11 A digitally programmable voltage source is formed by using a digital-to-analog converter to set the reference voltage for a voltage regulator.

to permit variation of output voltage in response to a command signal. With this control, high-speed automatic testing can be performed where supply or reference voltages must be changed, such as in power-supply rejection testing. Digital control of a regulator is illustrated with a basic circuit in Fig. 4.11. In this circuit the voltage reference is supplied by a digital-to-analog converter instead of a fixed reference element. By appropriate choice of digital input to the converter, a wide range of precise reference levels can be established for a great number of different regulator output levels. The precision of the output level is largely determined by the digital-to-analog converter, since it replaces the reference element that otherwise produces the dominant source of regulator output error. To maintain the inherent accuracy of the converter, it is generally desirable to bias it from a separate power supply. That supply should be used for the negative supply bias of the operational amplifier so that the amplifier can operate at voltages near ground when low output voltages are desired.

4.1.4 Extending regulator utility   Operational amplifiers can also be used in conjunction with completed voltage regulators to extend output capability. Specifically, operational amplifiers can split a single regulator output voltage into two for a dual supply, or a variable output control can be derived for use with fixed-voltage regulators. A simple means of converting a single-voltage power supply to a dual supply where power efficiency is not critical is that of Fig. 4.12. A voltage follower with an output current booster is referenced to the midpoint of a regulator output voltage in order to establish the appropriate third output terminal. That

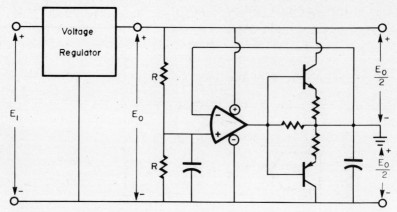

Fig. 4.12 A dual voltage regulator can be derived from a single output regulator by using a voltage follower to establish a common return.

output terminal voltage is midway between the original two, so it serves as a common return for the voltages maintained at the original terminals. Thus, the output supplies two opposite polarity voltages equal in magnitude to one-half the original regulator output voltage.

To assure conduction of currents returned to this derived common, the power booster from Fig. 1.10 is added to the amplifier output. Note that this power booster conducts the common current under the full voltage of one output; so power dissipation can be high. Where this dissipation is not acceptable, a second unregulated input supply can be derived to permit use of the circuit in Fig. 4.13.

The second input supply is used to power the new output as controlled by an added regulator circuit. Such an approach is most useful when a second supply voltage is needed to supply only a fraction of the power

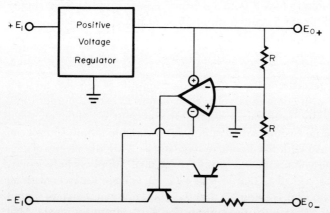

Fig. 4.13 A second regulator output voltage can be derived with a tracking regulator referenced from the original output voltage.

required of the original supply voltage. It is then more convenient to simply add the circuitry shown than to use a high-power dual supply. In a tracking regulator configuration, the added circuitry uses the original output voltage E_{0+} as a reference to set the new voltage E_{0-}. Only when the two voltages are equal in magnitude will the inverting amplifier input be held at zero voltage by the voltage divider, as required for the equilibrium state. Alternatively, the voltage divider can be unbalanced to establish a different magnitude for the negative output voltage.

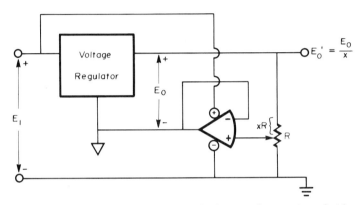

Fig. 4.14 Variable output voltage with a fixed-voltage regulator is achieved with a feedback amplifier that drives the common return of the regulator.

For a variable output voltage from a fixed-voltage regulator, an operational amplifier can be used to drive the common return of the regulator as in Fig. 4.14. By setting the voltage at the regulator common above the ground level, the regulator output voltage is similarly moved away from ground. The original output voltage E_0 will be maintained on the xR portion of the control potentiometer by the feedback action of the voltage follower. As a result, a current flows in the potentiometer to bias the noninverting amplifier input above ground level. This bias voltage is followed by the amplifier to raise the voltage at the regulator common terminal for a net output voltage of E_0/x. To do this, the amplifier must conduct the common return current of the regulator but not that from the load, which is still returned through the ground line. Note that the bias voltage supplied to the amplifier input by the potentiometer must be large enough to reach the common-mode input range of the amplifier and to set the amplifier output voltage above its saturation level.

4.2 Active Filters

With active filters, response poles and zeros are precisely and easily set through the use of operational amplifiers to isolate response-controlling elements from the loading effects of others. Numerous active filter configurations have been derived, and among them one of the most convenient is the state-variable form.[1] Described in this section are two simplifications of the state-variable filter that retain most of the response control features of the basic form. Also made possible by active filters is electronic control of filter characteristics, and means for digital control are presented for the basic filter functions.

4.2.1 Simplified state-variable configurations A major advantage of the state-variable active filter connection over simpler forms is the degree of freedom with which center frequency, midband gain, and selectivity can be set for a bandpass response. Each of these three characteristics can be separately determined because of the complete isolation of the individual filter elements by operational amplifiers. To achieve this isolation, three operational amplifiers are commonly required, as opposed to the single amplifier of less controllable forms. From these three operational amplifiers, three filter outputs are available, and they simultaneously provide low-pass, bandpass, and high-pass responses.[1]

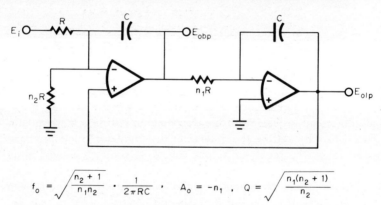

$$f_o = \sqrt{\frac{n_2 + 1}{n_1 n_2}} \cdot \frac{1}{2\pi RC} \quad , \quad A_o = -n_1 \quad , \quad Q = \sqrt{\frac{n_1(n_2 + 1)}{n_2}}$$

Fig. 4.15 The state-variable active filter connection can be simplified from three operational amplifiers to two through the use of the second input of one integrator to avoid the summing inverter commonly used.

Where only the bandpass or low-pass response is required, the state-variable active filter can be simplified to two-amplifier configurations for either inverting or noninverting operation. This is made possible by utilizing the second input of one filter amplifier for the summing and inverting normally performed by a separate amplifier. For inverting gain operation, this is accomplished with the circuit of Fig. 4.15. As shown, this structure retains the two integrators of the common circuit, but it eliminates the summing inverter through the use of the noninverting input of one amplifier. Retained by the simplified form are the two response poles and response zero required for the bandpass characteristic, and three degrees of freedom are provided by the three resistors to permit separate determination of bandpass gain, selectivity, and center frequency.

To set these characteristics, the midband gain is first fixed by n_1 and

$$A_o = -n_1$$

Then, selectivity is determined by the choice of n_2 by using the expression

$$Q = \sqrt{\frac{n_1(n_2 + 1)}{n_2}}$$

Last, center frequency is fixed by selection of R and C with the relationship

$$f_o = \frac{Q}{2\pi n_1 RC}$$

While the above characteristics can be separately determined, they are interrelated; so response adjustment does result in interaction. Variation of n_2 to adjust Q will alter f_o, and adjustment of f_o by variation of R effectively alters the n_1, n_2 ratios, disturbing the other characteristics. The bandpass response is obtained from the output labeled E_{obp}, and the low-pass response from the other.

Where noninverting or positive gain response is desired, an alternative two-amplifier state-variable active filter configuration is available. In this case, a differential integrator is used to avoid the need for an additional summing inverter, as in Fig. 4.16. For this circuit the bandpass gain is unity, and selectivity and center frequency are set by choice of the filter

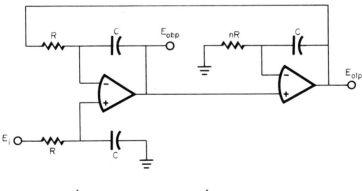

$$f_o = \frac{1}{2\pi RC\sqrt{n}}, \quad A_o = 1, \quad Q = \frac{1}{\sqrt{n}}$$

Fig. 4.16 A noninverting state-variable filter can be formed with two operational amplifiers through the use of a differential integrator that precludes the need for the normal summing inverter.

elements. First, selectivity is set by choice of the resistor ratio n for the expression

$$Q = \frac{1}{\sqrt{n}}$$

Then, center frequency is fixed by assigning values to R and C by using the expression

$$f_o = \frac{1}{2\pi \, RC \sqrt{n}}$$

As with the previous circuit, adjustments of Q and f_o interact.

In addition, adjustment is made more difficult by the need for close matching of the differential integrator elements. Each RC network of this integrator determines a response pole–zero pair, and a pole-zero cancellation is required to achieve the desired bandpass response. Cancellation occurs when the integrator RC networks match exactly. Otherwise, an unwanted pole-zero pair exists which is most notable in its disturbance of the bandpass gain from unity. Fortunately, this unwanted response pair can be removed without disturbing other characteristics through adjustment of the resistor connected to the input terminal. However, this then precludes variation of the matching network for adjustment of other response characteristics.

4.2.2 Digitally controlled active filters

In general, tunable active filters are adjusted manually using potentiometers. Digital electronic tuning provides opportunities for automatic adjustment, and such tuning can be achieved with a multiplying digital-to-analog converter (MDAC) in active filters. By multiplying the signal voltage impressed on a resistor, the resulting current is increased as though the resistance were divided by the same factor. As a result, changes in effective time constants can be achieved.[2] Analogous observations can be made considering multiplication of the signal voltages applied to capacitors. With this ability to control time constants, very rapid adjustments to filter characteristics can be made through variation in digital control voltages on multiplying digital-to-analog converters. This technique is described below for basic low-pass, high-pass, and bandpass active filter stages. Using these basic stages, or just their tuning techniques, more complex tunable filters can be formed.

The digitally tuned low-pass filter of Fig. 4.17 is simply an integrator with a second feedback loop through an MDAC. Since the magnitude of the second feedback signal is controlled by the MDAC, the effective time constant of the filter response can be controlled. This control function is defined by the signal E_r applied to the MDAC reference input. When

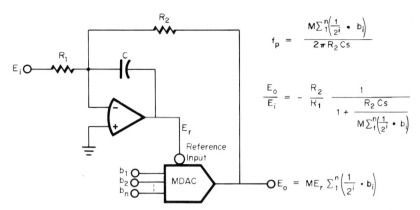

Fig. 4.17 A digitally controlled low-pass filter is formed by connecting a multiplying digital-to-analog converter in feedback with an integrator.

multiplied by the digitally weighted gain of the MDAC, it produces an output signal of

$$E_o = M E_r \sum_1^n \frac{1}{2^i} b_i$$

where M is the gain constant of the MDAC and b_i represents the various bit inputs.

By relating E_r to the input signal, the input-output response of the filter is defined as

$$E_o = \frac{-E_i R_2/R_1}{1 + R_2 Cs/M\sum_1^n \frac{1}{2^i} b_i}$$

From this expression, it is seen that low-frequency gain is not affected by the digital control, but the response pole frequency is directly related to the digital input by

$$f_p = \frac{M\sum_1^n \frac{1}{2^i} b_i}{2\pi R_2 Cs}$$

This response control is achieved only if the gain constant M is positive, to ensure correct feedback polarity. Also, the MDAC must be capable of accepting both signal polarities at its reference input if the input signals are bipolar. Note that switching transients, or *glitches*, accompanying bit changes at the MDAC input will appear at the filter output.

For a voltage-tunable high-pass filter, the inverse of the above circuit

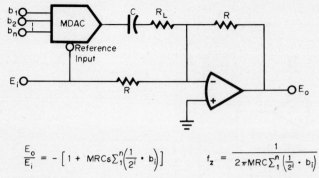

Fig. 4.18 For digital control of a high-pass filter response, a multiplying digital-to-analog converter is connected in series with the differentiator input of this circuit.

can be used. It is shown in Fig. 4.18 and consists of a differentiator-type circuit with an MDAC in the input signal path. Since the signal applied to the capacitor is multiplied, the effective capacitance is divided, and the differentiator gain term is decreased. The overall output signal is

$$E_o = -(1 + MRCs \, \Sigma_1^n \frac{1}{2^i} b_i) \, E_i \quad \text{for } R_L \ll R$$

From this high-pass function results a response zero that is inversely proportional to the MDAC response as expressed by

$$f_z = \frac{1}{2\pi \, MRC \, \Sigma_1^n \frac{1}{2^i} b_i}$$

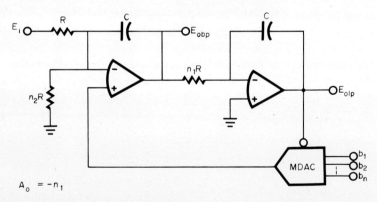

Fig. 4.19 Digital control of bandpass filter response is achieved with a multiplying digital-to-analog converter in the feedback path of the state-variable filter of Fig. 4.15.

As before, the MDAC gain constant must be positive, and MDAC switching transients are coupled to the filter output.

For digital control of a bandpass filter an MDAC is inserted in the feedback loop of the state-variable filter of Fig. 4.15. The result is the circuit in Fig. 4.19, which has both bandpass and low-pass outputs. For the bandpass output, center frequency and selectivity are controlled by the digital command signal as expressed in the figure. Both characteristics vary with the square root of the digital command, and they do so in a manner that leaves bandwidth constant at $1/n_1RC$. The midband gain is also independent of the digital control and remains at $-n_1$. In this state-variable filter case, some reduction in switching transient is provided by the filter, but such transients can be too fast to be removed unless the operational amplifiers have very wide bandwidths.

4.3 Frequency Multipliers

A test or reference signal having a frequency that is a multiple of that of a given signal can be derived by means of frequency multiplying techniques. For sinusoidal signals, frequency doubling can be performed with analog multipliers, and for triangle waves an absolute-value circuit performs the same task.[7-10] By cascading such frequency doublers, greater multiples of frequency can be attained. Means for multiplying the frequency of square waves are described in this section.

If square-wave frequency multiplication without precise duty cycle control is required, it can be performed with the circuit of Fig. 4.20. With this approach a second square wave is derived that is delayed in time with respect to the input signal as shown. This delayed square wave is compared with the original square wave by an exclusive OR gate which produces a high output state when one, but not both, square wave is positive. The result is a double frequency output.

To generate the delayed signal, an integrator and zero-crossing detector are connected in a feedback loop. Integration of the input square wave results in a triangle-wave output, and that output is made to cross zero by feedback from the zero-crossing detector. Feedback through the R_2C_2 low-pass filter develops a dc bias at the integrator input which then controls the dc level of the integrator output. Otherwise the dc output level would be increased to the integrator saturation level in the presence of any input signal of nonzero average value. With this feedback the input square wave can have both time and voltage dissymmetry without disturbing the circuit operation except for the associated output dissymmetry.

The range of frequencies that can be doubled by this circuit is determined by the selection of R_1, C_1, R_2, and C_2 and by the signal amplitude. When R_1 and C_1 are set, the integrator gain is determined, and this gain decreases with frequency. Thus, the amplitude of the integrator output

122 Designing with Operational Amplifiers

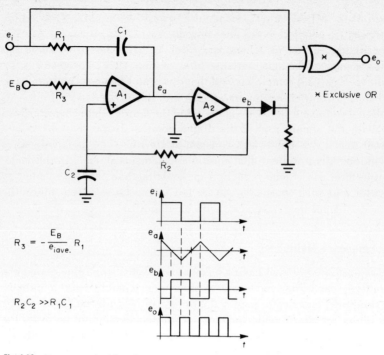

$$R_3 = -\frac{E_B}{e_{iave.}} R_1$$

$$R_2 C_2 \gg R_1 C_1$$

Fig. 4.20 Frequency doubling for a square-wave signal is achieved by generating a time-delayed square wave for comparison with the input signal by an exclusive OR gate.

signal will vary with frequency as well as with input signal magnitude. For circuit operation the integrator output signal must be less than the output saturation levels of A_1 but large enough to drive A_2 for rapid switching. This range of integrator output amplitudes places limits upon the amplitude and frequency of the input signal for circuit operation. Another limit is introduced by the filtering of R_2 and C_2. By making these components large in value, ripple is controlled in the feedback voltage applied to the integrator input. However, large values also limit the circuit response to changes in input signal average value. In any event, for adequate ripple control R_2 and C_2 must be chosen so that their time constant is many times that of R_1 and C_1.

With the circuit of Fig. 4.20 output dissymmetry with respect to time results from both time and voltage dissymmetry of the input signal, plus unwanted delay in the zero-crossing detector. Input signal time dissymmetry is transferred directly to the output signal. Both time and voltage dissymmetry in the input signal result in a nonzero average value that must be compensated for by the feedback from A_2. To provide this feedback, A_2 must produce an output dissymmetry with respect to time.

Some reduction in this effect can be achieved for input signals of relatively constant average value by adding the bias voltage E_B connected to R_3.

A more general-purpose frequency multiplier results from combining a frequency-to-voltage converter with a voltage-to-frequency converter as in Fig. 4.21. With this circuit, operation is insensitive to input signal wave shape, and frequency can be multiplied by any whole or fractional value. Output symmetry is not affected by either voltage or time dissymmetry in the input signal, and the input signal does not have to be a square wave. For the frequency-to-voltage conversion, amplifiers A_1, A_2, and A_3 provide a voltage E_{fi} by time-averaging the pulses created with each cycle.[2]

Amplifier A_1 operates as a comparator to convert the signal to a controlled-

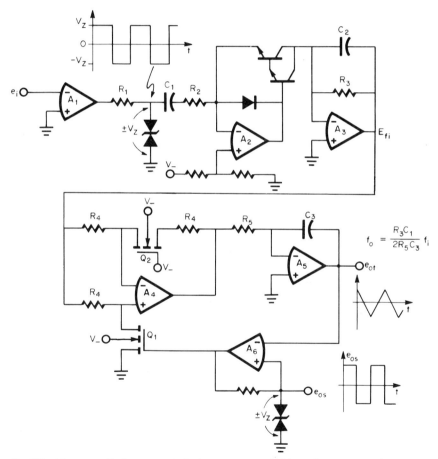

Fig. 4.21 More versatile frequency multiplication is provided by a frequency-to-voltage converter cascaded with a voltage-to-frequency converter.

amplitude square wave. If the input signal has unequal time intervals above and below the zero level, the comparator output will be similarly unsymmetrical, but this will not affect the output voltage developed. To improve the rise and fall times of an operational amplifier used for the comparator, its phase compensation should be removed.

Leaving the comparator, the signal is differentiated, rectified, and then averaged to produce a dc output. The differentiation and rectification are performed with A_2. From the rapid rises and falls of the comparator output voltage, differentiation capacitor C_1 produces current pulses that are supplied to either the transistors or the diode around A_2. Only one polarity of the current pulse is conducted by the transistors to the output amplifier A_3. This rectification action results in only one current pulse per cycle to A_3, rather than pulses with both the rise and fall of the signal. As a result, the time average of the pulses reaching A_3 is independent of the signal symmetry.

The averaging is performed by R_3 and C_2, and the average current in R_3 is determined by the change in C_1 charge that generated it and by the time between pulses. This is expressed by

$$\Delta Q = 2V_zC_1 = \bar{i} \Delta t \quad \text{for } R_2C_1 \ll \Delta t$$

Here the term Δt is the time between pulses, and this is the period of the signal, or the inverse of the frequency. Thus, the average voltage generated by the flow of $\bar{i}$ in R_3 is related to the input signal frequency by

$$E_{fi} = 2V_zR_3C_1f_i$$

At low frequencies, the accuracy of this frequency-to-dc conversion is determined primarily by the components denoted in the above expression. However, at higher frequencies the parasitic and stray capacitances of the differentiator-rectifier circuit introduce charge errors. At some high frequency the slewing-rate limit of A_1 will prevent it from completing its output swing, and very large errors will develop.

The voltage E_{fi} determines the output signal frequency by controlling the voltage-to-frequency converter formed with A_4, A_5, and A_6.[2] If the output of A_6 were connected to resistor R_5, deleting A_4 and its connecting circuitry, an integrator-comparator feedback loop would be formed like that of common square- and triangle-wave generators. In such a generator, the frequency is controlled by the amplitude of a square wave supplied to the integrator from the comparator. By inserting the circuitry of A_4 into the feedback loop of this generator, the generator frequency is controlled by the voltage supplied by A_4 to the integrator. That voltage will be a square wave with amplitude of plus and minus E_{fi} because of the switching of Q_1. The polarity of the gain provided by A_4 to E_{fi} reverses each time the integrator output reaches a comparator trip point and causes the comparator to reverse the state of switch Q_1. Reversing the state of this switch converts the amplifier configuration from that of an inverter to a follower,[2] and this reverses the polarity of the signal reaching the integrator input. With either gain polarity, E_{fi} controls the magnitude of the integrator input voltage and thereby controls frequency by the relation

$$f_o = \frac{E_{fi}}{4V_Z R_5 C_3}$$

Combining this expression with that previously noted for E_{fi} defines the frequency multiplication function of the circuit by

$$f_o = \frac{R_3 C_1}{2 R_5 C_3} f_i$$

REFERENCES

1. G. Tobey, J. Graeme, and L. Huelsman, *Operational Amplifiers: Design and Applications*, McGraw-Hill Book Company, New York, 1971.
2. J. Graeme, *Applications of Operational Amplifiers: Third-Generation Techniques*, McGraw-Hill Book Company, New York, 1973.
3. H. Jones, Reference Voltage Can Be Varied and the Optimum Zener Current Maintained, *Electron. Des.*, October 25, 1974.
4. R. Spencer, Inexpensive Power Supply Produces Zero Ripple Output, *Electronics*, November 8, 1973.
5. L. Dixon and R. Tapel, Designers Guide to: Switching Regulators, Part 1, *EDN*, October 20, 1974.
6. K. Christensen, Modified Current Limiter Circuit Ensures Turn-on of Power Supply with Constant Current Loads, *Electron. Des.*, November 8, 1973.
7. D. Sheingold, *Nonlinear Circuits Handbook*, Analog Devices, Inc., Norwood, Mass., 1974.
8. J. Hilburn and D. Johnson, *Manual of Active Filter Design*, McGraw-Hill Book Company, New York, 1973.
9. L. P. Huelsman, *Active Filters*, McGraw-Hill Book Company, New York, 1970.
10. E. R. Hnatek, *Design of Solid State Power Supplies*, Van Nostrand Reinhold Company, New York, 1971.

5

ABSOLUTE-VALUE CIRCUITS

Precision rectifiers or absolute-value circuits are widely used in ac voltmeters and other signal monitoring devices. Magnitude detection in most average-reading voltmeters begins with an absolute-value conversion. For these requirements high-frequency response is needed along with high accuracy. With operational amplifiers various absolute-value or precision rectifier circuits can be formed to readily achieve the desired highly accurate full-wave rectification. In these circuits, rectification is achieved without sacrificing a significant portion of the signal to forward-bias rectifying diodes. By connecting these diodes in the feedback loop of an operational amplifier, this signal loss is reduced by the high gain of the amplifier. With this connection the feedback drives the diodes into and out of conduction for only very small signal changes, permitting rectification of very small signals. Numerous circuits providing this precision rectification have been described.[1,5] Others are presented here which provide circuit simplifications, differential inputs, and improved response accuracies.

5.1 Single-Amplifier Configurations

While the more precise absolute-value circuits require two or three operational amplifiers for full-wave rectification, a number of single-

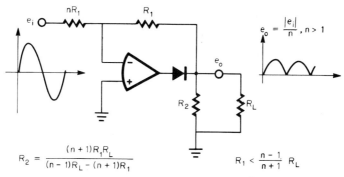

Fig. 5.1 Full-wave rectification with one operational amplifier is performed by a circuit that switches from an inverting amplifier to a voltage divider as the input signal polarity reverses.

amplifier configurations are available for less demanding requirements. Three such configurations are described in this section for voltage and current output applications. For voltage output, the simple circuit of Fig. 5.1 can be used, without undue loss of precision, through careful choice of resistor values.[2] Rectification is achieved when the circuit switches from an inverting amplifier to a voltage divider as the signal polarity reverses. When input signals are negative, the amplifier output is positive and is connected to the circuit output through forward-biasing of the diode. In this mode the circuit appears as an inverting amplifier with a gain of $-1/n$.

However, positive input signals drive the amplifier output negative, reverse-biasing the diode to disconnect the amplifier. Then, the input signal is passed to the output through a voltage divider. Since the resistive divider lacks low output resistance, output loading could greatly alter the output signal magnitude. But a load resistance can be considered as part of the divider to achieve the desired circuit attenuation. The attenuation required is equal to the magnitude of the inverting amplifier gain which controls the output for the opposite signal polarity. To set this attenuation at $1/n$, the magnitudes of R_1 and R_2 are selected considering the level of the load resistance R_L.

Several restrictions limit the utility of the circuit of Fig. 5.1: The signal is attenuated rather than amplified. Only constant impedance loads can be driven if loading error is to be avoided; so loads must either be resistive, or signal frequency must be constant with reactive loads. Additional error can be introduced by amplifier input current in the divider mode unless the operational amplifier maintains high input resistance under input overload. Frequency response and low-level rectification are limited by the amplifier slewing rate and dc input errors, respectively, as described in Sec. 5.4.

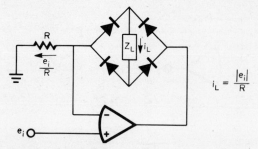

Fig. 5.2 Precise full-wave rectification with current output to a nongrounded load is provided by a diode bridge in an amplifier feedback loop.

Frequently, the signal desired from an absolute-value circuit is a current rather than a voltage, as for a meter drive in signal magnitude measurement. In these cases precise full-wave rectification is readily achieved with one operational amplifier. Where the circuit load can be floated, as in the case of meters, the load is connected in a feedback diode bridge as in Fig. 5.2. Rectification is performed by the diode bridge without normal bridge errors since diode voltage bias is supported by the operational amplifier. With this connection the diodes merely gate the amplifier feedback current to provide a unipolar load current. The feedback current is precisely controlled by the input signal voltage, because an equal voltage must be developed on the resistor at feedback equilibrium. Other benefits of this

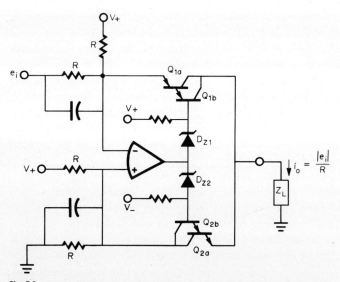

Fig. 5.3 A current output absolute-value circuit for driving grounded loads is formed with opposite polarity current-source feedback connections that alternately conduct current to the load.

simple circuit are high input impedance and freedom from the resistor matching commonly required for precision rectifiers. In this case, only one resistor controls the circuit gain, and that resistor may be varied for gain ranging.

Where a current output is acceptable but the load must be grounded, another single-amplifier absolute-value circuit is available that maintains high precision. Shown in Fig. 5.3, this circuit is formed with two current-source feedback connections[1] that alternately supply the loads as dictated by input signal polarity. One such feedback connection is through the Q_1 Darlington pair to the inverting amplifier input, and the other is made by the Q_2 pair to the noninverting input. Darlington pairs are used to limit signal current loss associated with base current, and their feedback resistors are bypassed to maintain frequency stability. To provide voltage bias for the transistors, both amplifier inputs are biased above ground potential by means of resistors connected to the positive supply. Additional biasing is supplied by the zener diodes to ensure that only one transistor pair conducts at any given time.

Input signal polarity determines which transistor pair will conduct. For positive input voltages the amplifier output will swing negative,

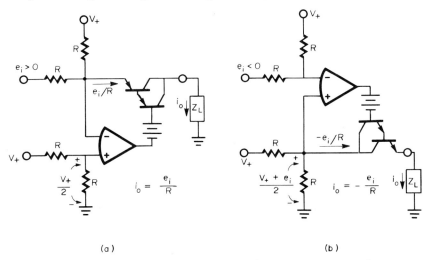

Fig. 5.4 For the two polarities of input signal the circuit of Fig. 5.3 uses opposite polarity current feedback paths to maintain unipolar output current.

forward-biasing the Q_1 pair and reverse-biasing the other. The result is a circuit equivalent to that of Fig. 5.4a. In this mode feedback will maintain the amplifier inputs at the voltage level established by the voltage divider on the noninverting input. Signal swing on the inverting input resistor must then be counteracted by feedback current through the conducting transistors. That current is supplied to the load and has a magnitude controlled by the input signal. Load voltage swing is limited by the input voltage bias and the Darlington saturation voltage.

When the input signal is negative, the amplifier output swings positive, turning the Q_1 pair off and the Q_2 pair on for a connection like that of Fig. 5.4b. Now the feedback is applied to the amplifier noninverting input, but the phase inversion of the transistors makes this feedback negative. Again, feedback will force the amplifier input voltages to be equal by conducting current through the feedback transistor. In this mode, the input voltages are controlled by the signal and power-supply connections to the inverting amplifier input, and the associated feedback current will have a magnitude controlled by the input signal. However, this current will have a polarity opposite that of the input signal; so the load receives the same polarity current as above for positive input voltages. Load voltage swing is limited by the input bias and the input signal swing.

5.2 Precision Absolute-Value Circuits

For greater accuracy in absolute-value conversion, two operational amplifiers are generally used in conjunction with a number of matched resistors and the rectifying diodes.[1] In the most common absolute-value configuration, resistor matching becomes a significant task if equal gain magnitudes are to be maintained for both positive and negative input signals. And circuit input resistance is generally set by one of the circuit resistors. Because of the need for several resistor matches, variable-gain precision rectifiers have been difficult to implement. Described in this section are precision rectifiers which circumvent these limitations.

To eliminate the multiple resistor matching requirement, an absolute-value circuit is available that requires matching of only one resistor pair as in Fig. 5.5. Only the two resistors labeled R must be matched to ensure

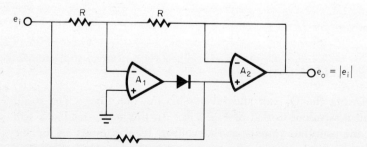

Fig. 5.5 Absolute-value conversion is performed with greater ease when only one resistor pair must be matched.

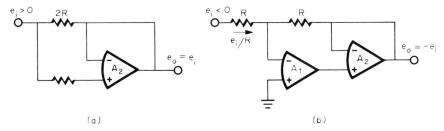

Fig. 5.6 To provide full-wave rectification, the circuit of Fig. 5.5 switches from a voltage follower to an inverter with a voltage-follower output.

equal gain magnitudes for the two signal polarities. Signal polarity determines whether the circuit performs as a voltage follower or as a voltage follower preceded by an inverter. Positive input signals drive the output of A_1 negative, reverse-biasing the diode to disconnect that output from A_2. As long as A_1 has high input resistance under input overload, this circuit mode is represented by Fig. 5.6a. Then, the signal presented to both inputs of A_2 equals the input signal; so A_2 acts as a voltage follower for positive unity gain.

When input signals are negative, the output of A_1 is driven positive to forward-bias the diode. Then the circuit operates as an inverter with a voltage-follower output, as in Fig. 5.6b. Considering A_2 as a voltage follower, it transfers the output voltage from A_1 to the circuit output without change. Thus, A_2 can be replaced by a short circuit in the ideal case, and the circuit is essentially an inverter. Circuit gain is then -1, or the inverse of the gain applied to positive input signals above. The result is a positive output independent of input signal polarity, as required for absolute-value conversion.

However, the circuit mode switching does pose two significant sources of error: First, the circuit input impedance varies dramatically from R to the common-mode input impedance of A_2 when input signal polarity reverses. As a result, any source impedance significant in comparison to R will result in unequal overall gains for the two polarities of input signal. Also, the bandwidth of this precision rectifier is restricted by potentially large voltage swings required of the output of A_1 when signal polarity reverses. This amplifier output voltage must swing from its negative saturation level to one forward diode voltage drop above ground at this polarity transition. Such transitions are limited by amplifier slewing rate, and major errors are developed as described in Sec. 5.4.2.

Most precision rectifier circuits like the above have low input impedances set by input summing resistors, and an additional buffer amplifier may often be needed. However, the circuit shown in Fig. 5.7 avoids the need for an additional amplifier because it presents an am-

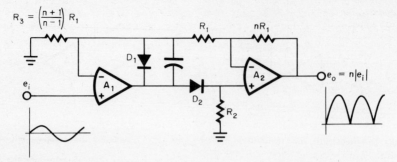

Fig. 5.7 High input impedance and simplified resistor matching is provided with this precision rectifier by avoiding summing resistors to the circuit input.

plifier common-mode input impedance, instead of summing resistors, to the input terminal.[3] This results in input resistances of typically 25 MΩ for bipolar transistor input amplifiers or 10^{12} Ω for FET input amplifiers.

Full-wave rectification is produced by diode switching that reverses the polarity of the net circuit gain when the polarity of the input signal reverses. In this way the polarity of the output signal is prevented from changing. This feature coupled with equal gain magnitudes for input signals of either polarity results in an absolute-value conversion.

Gain polarity is switched by the diodes as they alternate the connection of the output of A_1 between the two inputs of A_2. Positive input signals cause the output of A_1 to swing positive, reverse-biasing D_1 and forward-biasing D_2 for the connection represented in Fig. 5.8a. This connects the output of A_1 to the noninverting input of A_2; so A_2 provides a gain with positive polarity. Gain magnitude is controlled by the three feedback resistors that are multiples of R_1. Feedback forces the output of A_2 to that level which develops a voltage equaling e_i on R_3. For this positive signal case the associated gain is $e_o/e_i = n$. Both amplifiers are connected in a common feedback loop for the positive signal mode, and this may require additional phase compensation with the capacitor shown.

Negative input signals are amplified by a gain of opposite polarity. They cause the output of A_1 to swing negative, forward-biasing D_1 and reverse-biasing D_2, as represented by Fig. 5.8b. Now A_1 drives the inverting rather than the noninverting input of A_2. The noninverting input of A_2 is connected to ground through R_2. In this configuration A_2 is connected as an inverting amplifier and provides a negative gain to the signal supplied by A_1. With its feedback shorted by D_1, A_1 performs as a voltage follower and supplies the A_2 inverting amplifier with a signal equaling e_i. The result is an overall circuit gain of $-n$. Thus, the circuit gain switched from $+n$ for positive signals to $-n$ for negative signals, as desired for full-wave rectification.

A number of factors limit the performance of the circuit of Fig. 5.7, including resistor mismatch and the input offset voltages, input bias currents, slewing-rate limits, and gains of the operational amplifiers. The input offset voltages and input bias currents of the operational amplifiers offset the transfer response as described in Sec. 5.4.1. The two errors are removed by first nulling the offset voltage of A_1 to remove the transfer response offset and then nulling A_2 to remove circuit output offset voltage. Because of interaction of the two nulls, this procedure must generally be repeated. High-frequency rectification is limited by a dead band around zero resulting from limited slewing rate and gain in A_1 and from diode capacitances, as described in Sec. 5.4.2.

Any deviation in resistor match from the ratios indicated will produce gain error which in some cases will make the gain magnitudes different for the two polarities of input signal. This gain error is removed by first adjusting the gain for negative signals by trimming R_1 or nR_1. Gain for positive signals can then be matched to the latter by adjusting the resistor R_3. Prior to these gain trims, the dead-band nulling described above may be required, because that error can also produce unequal outputs for equal positive and negative input signals.

While a number of circuit realizations have been developed for absolute-value conversion, variation of their gains requires adjustment of more than one resistor in all but the most complex of these circuits.[1,3] Variable gain is needed for range control in more common precision rectifier applications, such as amplitude detection in ac voltmeters. This control is commonly achieved with a separate input amplifier that also serves as an input buffer. However, with the precision rectifier of Fig. 5.9 shown, variable gain is achieved without a separate gain control amplifier.[4] Gain is controlled by a single resistor, which can be a potentiometer or multiple-tap network. In addition, this circuit has high input impedance

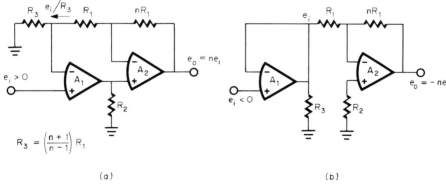

Fig. 5.8 To perform rectification, the circuit of Fig. 5.7 switches from a compound noninverting amplifier to a voltage follower and an inverting amplifier.

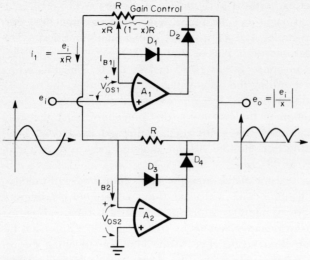

Fig. 5.9 Precision rectification with variable gain is provided by this high input impedance circuit using only two matched resistors.

without an input buffer and requires only one resistance match. These added qualities alone are rare in precision rectifier circuits.[1] Provided by this circuit is a gain range from almost unity to several thousand for a signal range of 1 mV to 10 V. Full-scale error can be reduced to 0.03 percent, and even for millivolt signals the error can be held to a few percent.

Rectification results from a switch in circuit gain polarity with the signal polarity reversal. Gain polarity is reversed by signal-induced switching of the feedback diodes. When the diodes switch, the signal path to the output changes from a noninverting amplifier to a voltage follower and an inverting amplifier. Positive input signals produce a positive current i_1 that drives D_2 and D_3 on and D_1 and D_4 off, resulting in a circuit connection represented by Fig. 5.10a. This connects A_1 as a noninverting amplifier with a gain of $1/x$. In this mode A_2 merely serves as a ground return for the resistance xR. Otherwise, A_2 is disconnected from the circuit by the reverse-biased D_4; so the circuit output is controlled by A_1 alone to be $e_o = e_i/x$.

When the input signal swings negative, so does the current i_1 to switch off D_2 and D_3 and turn on D_1 and D_4 for a connection like that of Fig. 5.10b. Now the output of A_2 rather than A_1 is connected to the circuit output to control e_o. Amplifier A_1 merely serves to maintain a signal equal to e_i at its own inverting input. In doing so, it develops this signal on the resistance xR. That resistance now acts as the input resistor to an inverting amplifier formed with A_2. With a gain of $-1/x$ this inverting amplifier develops $e_o = -e_i/x$, which is the negative of that produced by positive signals. Since the polarity of the gain switches with that of the input signal, the output signal is always positive, and

$$e_o = \left| \frac{e_i}{x} \right|$$

Gain can be varied from near unity to several thousand to accommodate a wide range of signal levels. To ensure continually equal gain for positive and negative signals, it is only necessary to match the resistor shown to the total potentiometer resistance. Operational amplifier gain error will directly affect circuit gain, but this effect is essentially identical for positive and negative signals.

Otherwise, circuit accuracy is dependent upon the noises, dc errors, and ac responses of the amplifiers. Noise is not generally a major source of error in the practical signal range of 1 mV to 10 V, so long as the resistance levels are low enough to limit the effects of amplifier input noise currents. The dc input offset voltages and bias currents of the amplifiers offset the diode switching from the input signal zero crossing with an equivalent input error of $V_{OS1} - V_{OS2} + I_B x R_1$. Because of the switching-point offset, small signals will not be rectified. To extend low-level operation, the operational amplifier offset voltages are nulled, and the amplifiers are chosen for suitably low input bias currents. A circuit output offset is also produced by the flow of amplifier input currents through the feedback resistances. This offset cannot be removed by the opera-

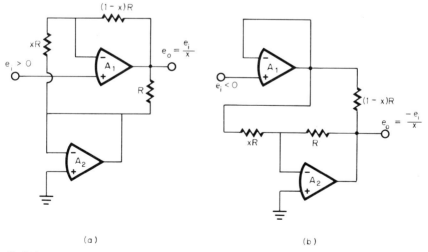

Fig. 5.10 When input signal polarity reverses, the circuit of Fig. 5.9 switches from a noninverting amplifier to a voltage follower and an inverting amplifier.

tional amplifier null controls without again offsetting the diode switching, but it is readily minimized by choice of suitable amplifiers and resistor levels. High-frequency performance is limited by the speed with which the amplifier outputs can turn off one rectifying diode and turn on another as described in Sec. 5.4.2.

5.3 Differential Input Absolute-Value Circuits

Precision rectifiers are widely used for amplitude detection in ac voltmeters because they provide rectification of signals from millivolts to volts. However, the numerous circuit realizations of these rectifiers[1,5] lack the differential inputs desirable for voltmeters. To provide differential inputs, the precision rectifier can be preceded by an instrumentation amplifier or by an operational amplifier connected in the difference amplifier configuration. A more economic solution is to configure the precision rectifier with differential inputs, as done with the four circuits described below. Described are circuits having high or low input impedance and for grounded or floating loads.[6]

The simplest circuit requires only two operational amplifiers and four matched resistors, and it is suited for applications where lower input impedance and frequency response are adequate. One of the amplifiers performs as a voltage-to-current converter and the other as a rectifying current-to-voltage converter[1] as shown in Fig. 5.11. By converting the signal voltage to a current, accurate rectification is simplified, since the voltage drops of rectifying diodes do not introduce error to a signal current. This conversion is performed by the differential input voltage-controlled current source from Fig. 2.15 and formed with A_1 and the R_1, R_2 resistors. Input signal e_2 supplies current directly to the rectifier formed with A_2, but this current is altered by the voltage e_L developed on this loading

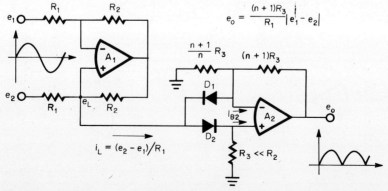

Fig. 5.11 Precision full-wave rectification of a differential voltage is achieved by transforming it to a current that is rectified then reconverted to a voltage.

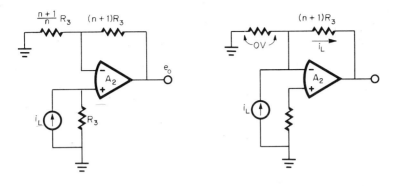

(a) $i_L > 0$, $e_o = (n + 1)R_3 i_L$ (b) $i_L < 0$, $e_o = -(n + 1)R_3 i_L$

Fig. 5.12 Rectification with the circuit of Fig. 5.11 results because the diodes switch opposite polarity currents to opposite amplifier inputs.

circuit. Additional current is supplied to the rectifier circuit through the R_2 positive feedback path. Controlling this current is the voltage developed by e_1 and e_L at the output of A_1. Fortunately, the effect of e_L on this additional current opposes that on the current supplied by e_2, and the two effects can be made to cancel. Cancellation occurs when the positive and negative feedback networks are matched as shown. Then the current supplied to the rectifier is independent of the voltage it develops on the rectifier. Supplied to the rectifier is a current

$$i_L = \frac{e_2 - e_1}{R_1}$$

To rectify this current, it is diode-gated to whichever input of A_2 will result in a positive output voltage, as illustrated in Fig. 5.12. When the current is positive, D_2 is on, as represented in Fig. 5.12a. The voltage developed on R_3 is amplified by a gain of $n + 1$, resulting in an output voltage of

$$e_o = (n + 1)R_3 i_L \quad \text{for } i_L > 0$$

Negative currents are conducted by D_1 through the $(n + 1)R_3$ feedback resistor as in Fig. 5.12b. None of the current is diverted by the other feedback resistor because the inverting input of A_2 is now at essentially ground potential. As a result, the output voltage developed is

$$e_o = -(n + 1)R_3 i_L \quad \text{for } i_L < 0$$

Note that gain is now the negative of that supplied to positive currents; so

$$e_o = \frac{(n + 1)R_3}{R_1} |e_1 - e_2|$$

Low-frequency performance of this precision rectifier is limited by the current-source resistors, by their mismatch, and by the dc errors of the operational amplifiers. Circuit input resistance varies with the load resistance on the current source and is as low as R_1 from either input to ground. Careful matching of the R_1, R_2 resistor sets is necessary to maintain high current-source output resistance and high common-mode rejection. For perfect resistor matching the output resistance approaches R_1 times the common-mode rejection ratio of the operational amplifier, and the common-mode rejection of the current source then equals that of the amplifier. The input offset voltage and input offset current of A_1, V_{OS1}, and I_{OS1} produce a current-source error of $I_{OS1} + V_{OS1}/R_1$. Generally, the total error current can be compensated for by adjusting the amplifier null, unless R_1 is small.

Similarly, the input offset voltage of A_2 is readily removed. However, for this amplifier the effects of input bias currents I_{B2} cannot be removed by the conventional null control. These currents offset the diode switching point from the signal zero crossing, causing rectification to cease at low signal levels. Before D_2 will turn off, the current source must overcome the forward-biasing effect of I_{B2}. Similarly, D_1 is held off by the other input current of A_2. To reduce this switching offset, a compensating current offset is developed in the current-source output. The null control of A_1 is used to adjust this compensation while observing the rectification of a small signal.

As with most precision rectifiers, the high-frequency performance of the circuit of Fig. 5.11 is limited by the ability of A_1 to rapidly drive the diodes on and off at the zero crossing. At this transition, the output of the current source must swing a voltage equal to two diode drops, and during that swing the output signal is in error. The time required for switching is determined by the bandwidth or slewing rate of A_1, as described in Sec. 5.4.2.

For the circuit of Fig. 5.11, the switching time is greatly increased by the requirement that the current source be loaded by a resistance much less than that of R_2. With $R_3 \ll R_2$, the output swing of A_1 is greatly attenuated before reaching the diodes. This requires the amplifier output to swing many times the diode switching voltage, and a similar increase in transition time results. For this reason, ac error is greater with this configuration.

Where high input impedance is required at only one of the differential inputs, the precision rectifier of Fig. 5.13 can be used. Here the impedance presented to one input is the high common-mode input impedance of A_1, although that at the other input remains set by a resistor R_1. Also improved over the previous circuit is bandwidth, because the full operational amplifier output drives the rectifying diodes. The time re-

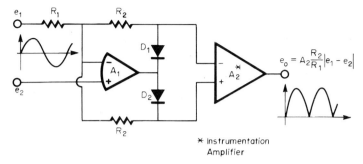

Fig. 5.13 High input impedance at one of the differential inputs of a precision rectifier results with an operational amplifier whose output is switched between the inputs of an instrumentation amplifier.

quired for switching the diodes remains the principal bandwidth restriction, as described with the previous circuit. However, the operational amplifier is only required to swing a voltage equal to two forward diode drops for diode switching, and the precision rectifier bandwidth is significantly extended.

Rectification results as the diodes switch the operational amplifier output from one instrumentation amplifier input to the other. This reverses the polarity of the gain provided by the instrumentation amplifier, whenever signal polarity changes, so that the circuit output signal will always be positive. When the differential input signal is positive, the output of A_1 swings negative to forward-bias D_1 and reverse-bias D_2. Then, the output of the operational amplifier is connected to the inverting input of A_2, and the signal supplied to that input is $e_2 + (e_2 - e_1)R_2/R_1$. The instrumentation amplifier will amplify the difference between this signal and the one at its noninverting input. That input is connected through a resistor to the inverting input of A_1, which follows the signal e_2. Essentially the same signal reaches A_2 because the high impedance input of A_2 conducts very little current through the resistance. The result is an output signal of

$$e_o = A_2 \frac{R_2}{R_1} (e_1 - e_2) \quad \text{for } e_1 - e_2 > 0$$

Negative differential input signals reverse the diode conduction states and reverse the signal connections to the instrumentation amplifier. This switches the polarity of the gain supplied to the two signals above to maintain a positive output. With the output signal always positive,

$$e_o = A_2 \frac{R_2}{R_1} |e_1 - e_2|$$

Other performance of this precision rectifier is primarily determined by resistor matching, the normal errors of the amplifiers, and the switching offset common to precision rectifiers. From mismatch in the R_2 resistors, unequal gain magnitudes result for positive and negative signals. Input signal range is limited by the minimum gain provided e_2 by the operational amplifier. That gain is always greater than 1; so the signal range for this input is often limited to less than that of the operational amplifier output. Common-mode rejection ratio is that of the operational amplifier less R_1/R_2 times that of the instrumentation amplifier. Some reduction in common-mode rejection results from the unequal source resistances presented to the inputs of the operational amplifier and, similarly, to the inputs of the instrumentation amplifier.

The input offset voltages and input bias currents of the two amplifiers produce an output offset, as in any amplifier circuit, but an offset is also introduced to the diode switching threshold, as described for the previous circuit. Error currents in the feedback paths of A_1 bias the diodes on or off when the differential input signal is zero. For the circuit of Fig. 5.13 such error currents result from the input offset voltage of A_1 and from the input bias currents of both amplifiers. No rectification is achieved until the input signal overcomes the error currents, resulting in a loss of rectification at low signal levels. To extend low-level operation, the error currents can be largely compensated for by adjustment of the offset control of A_1. While this may increase the circuit output offset, that offset can be removed using the null control of A_2.

High impedance at both inputs of a differential input precision rectifier can be achieved simply where a current output into a floating load is suitable. These characteristics are provided by two operational amplifiers and without the need for resistor matching by the circuit of Fig. 5.14.

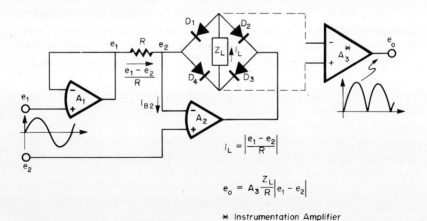

$$i_L = \left| \frac{e_1 - e_2}{R} \right|$$

$$e_o = A_3 \frac{Z_L}{R} \left| e_1 - e_2 \right|$$

* Instrumentation Amplifier

Fig. 5.14 High input impedance for a differential input precision rectifier is provided by two operational amplifiers that produce a current output convertible to a voltage with an instrumentation amplifier.

Both inputs present the high common-mode input impedance of an operational amplifier. Rectification is performed by the diode bridge as it routes the feedback current of A_2. That current is precisely controlled by the two amplifiers because their feedbacks determine the voltage on R. Each amplifier forces the signal at its noninverting input to follow its respective input signal e_1 or e_2. That fixes the voltage across R at $e_1 - e_2$ for a load current of

$$i_L = \left| \frac{e_1 - e_2}{R} \right|$$

Alternatively, a ground-referenced voltage output can be produced using the instrumentation amplifier shown as A_3 in Fig. 5.14. Then

$$e_o = A_3 \frac{Z_L}{R} |e_1 - e_2|$$

This approach does require the same amplifiers as a conventional precision rectifier preceded by an instrumentation amplifier, but resistor matching is less critical with the circuit of Fig. 5.14. Ratio error between R and Z_L affects gains for positive and negative signals equally, which is not the case with common precision rectifier resistor mismatch.

Circuit limitations include those of conventional follower-connected operational amplifiers, the diode switching offset, and the speed limitation of diode switching. As described for the previous circuit of Fig. 5.11, amplifier input offset voltages and bias currents offset the diode switching, preventing rectification of small signals. Switching offset for Fig. 5.14 results from a feedback error current of $I_{B2} + \Delta V_{OS}/R$, where ΔV_{OS} is the difference between the operational amplifier input offset voltages. Removal of the feedback error current is accomplished by adjustment of the operational amplifier offset controls to permit rectification of millivolt level signals.

Limitations to ac performance differ from those of conventional voltage followers in bandwidth and common-mode rejection. The precision rectifier bandwidth is restricted by the speed with which A_2 can switch its feedback diodes in the same manner as described with the previous circuits. Common-mode rejection for the precision rectifier is, however, often improved beyond those of its operational amplifiers. The common-mode errors of A_1 and A_2 tend to cancel if they are the same amplifier type.

Another differential input precision rectifier circuit provides high input impedance and a ground-referenced voltage output using three operational amplifiers and four matched resistors as in Fig. 5.15. This circuit consists of a differential input, rectifying, controlled current source[7] formed with A_1 and A_2, and a level-shifting current-to-voltage converter formed with A_3. To provide rectification, the current source switches its feedback current between Q_1 and Q_2. As with the preceding circuit, this feedback current is determined by the voltage established on resistor R. Feedback forces the ends of R to follow e_1 and e_2 for a feedback current i_f of $(e_2 - e_1)/R$. Note that this current flows out of the A_2 feedback path and into that of A_1; so the two feedback currents are equal but opposite in polarity. For a given polarity differential input signal the appropriate polarity feedback current is conducted to the output amplifier, and the other feedback current is absorbed by an amplifier.

When $e_1 < e_2$, i_f is positive and will be conducted by Q_1 and D_2. No current will then flow in Q_2, so long as enhancement-mode MOSFETs are used, and the circuit is then represented by Fig. 5.16. The conducting MOSFET supplies current to the inverting input of A_3. None of this signal current can flow in the R_1 resistor connected to Q_1 since the current in this resistor is fixed by the bias at the other input of A_3. To maintain equal voltages at its inputs, A_3 will conduct all the signal current through its R_2 feedback resistor. This produces an output voltage of $-i_f R_2$. For $e_1 > e_2$ a current of $-i_f$ will flow through Q_2 to the output amplifier, and $e_o = i_f R_2$. In this way only positive signal currents are supplied to A_3, and the output voltage will be

$$e_o = -\frac{R_2}{R}|e_1 - e_2|$$

No output voltage is developed from the negative supply connection to the R_1 resistors because this appears as a common-mode bias. That bias

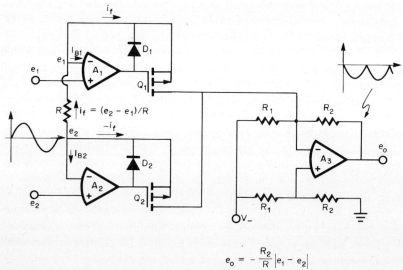

Fig. 5.15 A rectifying current source and a current-to-voltage converter form a differential input precision rectifier with three operational amplifiers.

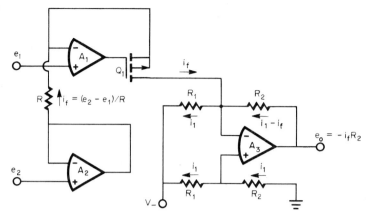

Fig. 5.16 For $e_2 > e_1$ the circuit of Fig. 5.15 supplies current to the output amplifier through Q_1 only.

holds the inputs of A_3 at a low voltage to permit signal voltage swing on the MOSFETs.

Factors limiting the accuracy of the circuit of Fig. 5.15 are similar to those that affect the basic operational amplifier connections resembled by parts of the circuit, plus the switching offset and bandwidth limitations common to precision rectifiers. Errors typical of voltage followers[5] affect the operation of A_1 and A_2, and the usual difference amplifier errors[5] are introduced by A_3. Circuit switching is offset by errors of $RI_{B1} + \Delta V_{OS}$ or $RI_{B2} + \Delta V_{OS}$, where ΔV_{OS} is the difference between the offset voltages of A_1 and A_2. Once again, the error currents are reduced by the amplifier offset controls. Bandwidth is more restricted for this circuit than for the preceding two, because A_1 and A_2 must swing greater voltages at the switching transition. Each amplifier must then swing a voltage equal to a forward diode drop plus the threshold voltage of the MOSFETs. Common-mode rejection is often improved beyond that of the input amplifiers, as with the last circuit, since the common-mode errors of A_1 and A_2 tend to cancel.

5.4 Absolute-Value Circuit Response Improvements

Both the ac and dc errors of operational amplifiers affect absolute-value circuits or precision rectifiers in manners quite different from most operational amplifier applications. The differences arise in those applications that employ diode switching within amplifier feedback loops. As a result of the diode switching, the offset nulling approach employed with precision rectifiers differs from those normally followed, and the bandwidth achieved with precision rectifiers is far lower than the basic bandwidth of the amplifier. Described in this section are techniques for removing the dc errors and extending the bandwidth of precision rectifiers.

5.4.1 Removing dc errors

By means of more judicious amplifier offset nulling,[8] precision rectifier operation can be extended to signals as low as 0.2 mV instead of a more common 5 mV limitation. Offset nulling is routinely used to reduce errors in operational amplifier circuits, but straightforward nulling can degrade, rather than improve, precision rectifier performance. In these circuits, rectification is achieved without sacrificing a significant portion of the signal to forward-bias rectifying diodes. Since the diodes are in the feedback loop of an amplifier, this signal loss is reduced by the high gain of the amplifier. The feedback drives the diodes into and out of conduction for only very small input signal changes, permitting rectification of millivolt level signals.

However, this low signal capability is not fully realized unless the unique offset nulling requirements of precision rectifiers are considered. To describe these requirements, the basic precision rectifier of Fig. 5.17 is used, but analogous requirements apply to other configurations. For this circuit straightforward null of A_1 does not result in the appropriate zeroing. Nulling the offset of the operational amplifiers does not cause the feedback diodes to switch at the signal zero crossing. These diodes switch when the polarity of the feedback current reverses and not necessarily when the input or output voltage is zero. Even when the offset of A_1 is zero, several sources of feedback current shift the switching point of the feedback diodes. Because of this switching point offset, rectification of low-level signals is prevented. No signal will be rectified unless it can overcome the feedback error current, and this typically requires 5 mV or greater signals.

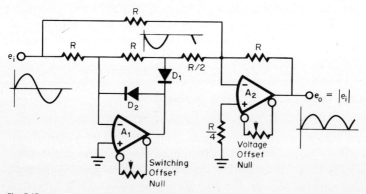

Fig. 5.17 Precision rectifier operation at low signal levels requires offset nulling of A_1 to remove switching offset only followed by nulling of A_2 to remove output offset voltage.

Precision rectification can be extended to a lower signal amplitude limit of about 0.2 mV by a special nulling procedure. First the input amplifier is offset-adjusted to cause the diodes to switch when the input signal crosses zero. This should be done with a small, low-frequency test signal such as 10 mV at 10 Hz. While observing the anode of D_1, the offset control of A_1 is varied until the peak-to-peak swing of the signal observed is exactly one-half that of the input signal. Then the overall circuit offset is removed by selecting a resistor to connect from either positive or negative supply to the summing junction of A_2. Alternatively, the offset may be removed using the offset control of A_2, but this will require iterative readjustment of the switching and voltage offsets, because the two interact.

5.4.2 Extending absolute-value conversion bandwidth The frequency response of common absolute-value circuits is unusually limited. It is far less than the frequency responses of the associated operational amplifiers because of the time required by the amplifiers to switch the rectifying diodes. With a typical operational amplifier, the usable full-power response and the small-signal bandwidth are reduced an order of magnitude in precision rectifier configurations. By means of the circuits to be described,[9] the precision rectifier full-power response can be boosted even above that of the operational amplifier, and its small-signal bandwidth can be largely restored.

The source of the precision rectifier bandwidth limitation can be seen by considering the common half-wave precision rectifier of Fig. 5.18.

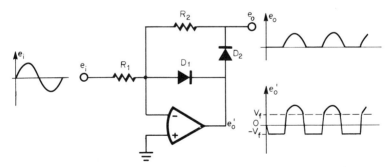

Fig. 5.18 To perform precision rectification, the operational amplifier must drive one diode on and the other off each time the signal polarity changes.

Most full-wave rectifiers make use of this circuit or circuits having analogous diode switching operation. High-frequency performance with most such circuits is limited by the speed with which the amplifier output e_o' can turn off one rectifying diode and turn on the other. While the first diode is being turned off, signal is passed with the wrong polarity; and while the second diode is turning on, no signal is passed. For no error during this transition it would have to be instantaneous. However, the slewing rate and gain-bandwidth product of the operational amplifier limit the speed with which the output of the amplifier can swing the required voltage equal to two diode voltage drops $2V_f$. If the input signal e_i is small, the rate of change of the amplifier output voltage will equal the rate of change of the input signal multiplied by the open-loop gain of the amplifier at the signal frequency $A(f_i)$. The transition time will then be the time required for the input signal to transverse a voltage of $2V_f/A(f_i)$. For larger signals, the rate of change of the amplifier output voltage is limited to its slewing-rate limit S_r, and the transition time will be $2V_f/S_r$. Since the ideal transition time would be zero, the response limitations imposed during this time by $A(f_i)$ and S_r are far more serious than those imposed in simply following the signal. This results in response limits at much lower frequencies for absolute-value conversion than encountered in simple amplifier applications.

To boost the speed with which the amplifier output drives the diodes, this output drive signal could be amplified. However, addition of gain in feedback circuits is accompanied by the need for added, stabilizing phase compensation, and the resulting gain-bandwidth product is not

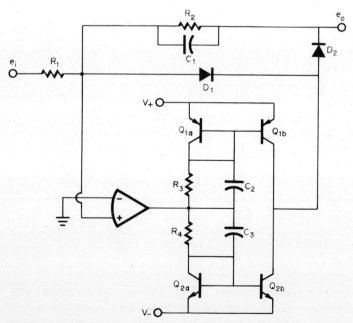

Fig. 5.19 Speed boosting gain is selectively added only during the switching transition by a gain stage with high output impedance.

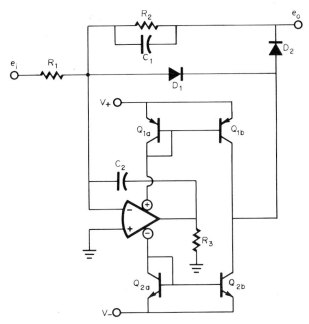

Fig. 5.20 Precision rectifier full-power response can be made even greater than that of the operational amplifier used if an added gain stage is driven from the amplifier supply current drains.

generally increased. Fortunately, the feedback loop is open during the switching transition; so the added phase compensation is not needed when the added gain is desired. If gain can be selectively added only during the open-loop switching transition, then switching speed will be boosted, and feedback stability following switching will be maintained.

The speed boosting operation described above is achieved with the circuit of Fig. 5.19. From the stage formed with Q_1 amd Q_2, a gain of several hundred is added during the switching transition. During this transition, both D_1 and D_2 are off and do not shunt the output of the added stage. Following the transition, one of the diodes conducts heavily, shunting the high output impedance of the stage and dropping its gain to less than unity. Thus, the high output impedance of the added stage ensures that gain is added only during the switching transition. The result is an increase in the full-power response of the precision rectifier to essentially that of the amplifier in this case.

Circuit elements for Fig. 5.19 are chosen to maintain low impedances for minimum RC delays and to provide transistor matching in the added stage. Resistors are kept as low as practical to speed charging of stray and parasitic capacitances. In particular, the feedback resistor R_2 should be small to hasten the turnoff discharge of the capacitance of D_2. For the same reason capacitance loading on the output terminal should be avoided. Accurate matching of like transistors can be achieved with monolithic pairs. Alternatively, unmatched transistors can be used if emitter degeneration resistors are added to stabilize biases.

To boost precision rectifier full-power response above that of the operational amplifier used, the added stage is driven from the power-supply current drains of the amplifier as in Fig. 5.20 and previously described with Fig. 1.13 in Sec. 1.4. With this circuit the output swing required of the operational amplifier is greatly reduced. This lower output swing does not reach the amplifier slew-rate limitation until a much higher frequency than the full-power response. A small R_3 load on the amplifier draws rated output current from the amplifier for only a small output voltage swing. Current drawn from the amplifier output must be supplied through Q_{1a} or Q_{2a}. This produces a matching current in Q_{1b} or Q_{2b}, which then drives the switching feedback impedance in the same way as the last circuit.

Circuit elements are chosen as described for Fig. 5.19 except for C_2 and R_3. Improved frequency stability is provided by C_2. The amplifier load resistor R_3 is chosen low enough to limit amplifier voltage swing requirements but not so low as to excessively lower amplifier gain.

REFERENCES

1. J. Graeme, *Applications of Operational Amplifiers: Third-Generation Techniques*, McGraw-Hill Book Company, New York, 1973.
2. R. Wincentsen, Absolute-Value Circuit Uses Only Five Parts, *EDN*, November 1, 1972.
3. J. Graeme, Full-Wave Rectifier Needs Only Three Matched Resistors, *Electronics*, August 8, 1974.
4. J. Graeme, Rectifying Wide-Range Signals with Precision Variable Gain, *Electronics*, December 12, 1974.
5. G. Tobey, J. Graeme, and L. Huelsman, *Operational Amplifiers: Design and Applications*, McGraw-Hill Book Company, New York, 1971.
6. J. Graeme, Measure Differential AC Signals with Precision Rectifiers, *EDN*, January 20, 1975.
7. D. Sheingold, *Nonlinear Circuits Handbook*, Analog Devices, 1974.
8. J. Graeme, Extend Precision Rectifiers to Very Low Levels, *EDN*, February 5, 1974.
9. J. Graeme, Boost Precision Rectifier BW above That of Op Amp Used, *EDN*, July 5, 1974.

6
SIGNAL GENERATORS

With the precision provided by high feedback loop gain, operational amplifiers generate highly accurate waveforms, and operational amplifier versatility permits generation of a great variety of waveforms. Described in this chapter are circuits that produce sine waves, square waves, triangle waves, pulse trains, ramp waves, staircase waveforms, and timed duration pulses. For each waveform circuit, alternatives are illustrated that offer varying degrees of simplicity, accuracy, and control.

6.1 Wien-Bridge Oscillator

One of the simplest oscillator configurations for generating sine waves of precise frequency is a Wien-bridge oscillator.[1] Operational amplifier realizations of this oscillator simplify the isolation of frequency-determining elements and the stabilization of the critical oscillator gain. However, such realizations commonly require two power-supply voltages and remain limited by the amplitude control and frequency variation difficulties of Wien-bridge oscillators. Each of these limitations is addressed by the circuits of this section. Described are operational amplifier oscillators for single-supply applications, with input amplitude control and with single-resistor control of frequency.

To permit single power-supply operation, a common-mode bias is estab-

150   Designing with Operational Amplifiers

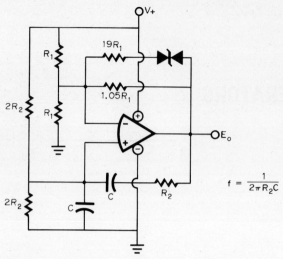

Fig. 6.1 Single-supply operation of an operational amplifier Wien-bridge oscillator is made possible by replacing frequency- and gain-setting resistors with biasing voltage dividers.

lished at the inputs of the Wien-bridge amplifier as in Fig. 6.1. This bias is supplied by the two voltage dividers connected from the power-supply voltage to ground, and the divider resistors also serve as gain- and frequency-setting elements. Setting the frequency are the R_2, $2R_2$, and C Wien-bridge elements. These elements supply positive feedback around the amplifier to induce oscillation, and oscillation results at the frequency $f = 1/2\pi R_2 C$, where the positive feedback peaks. A feedback peak results as the series capacitor increases feedback with frequency while the parallel capacitor decreases it. For equal resistances and capacitances in the Wien bridge, as shown, the peak feedback factor is $1/3$ (Ref. 2). Then, for a gain of 3 through the amplifier, the gain around the positive feedback loop is unity and oscillation results. The amplifier gain is set by the feedback to its inverting input.

Deviations from unity in the gain around the positive feedback loop cause the oscillation amplitude to diverge or converge with time. Extremely precise gain setting would be required for amplitude stability without AGC (automatic gain control). With AGC the initial gain is set slightly high to initiate oscillation; however, the greater this gain the greater the distortion. As the signal amplitude approaches the desired peak level, AGC feedback begins to reduce gain and stops the signal increase. Without AGC the signal amplitude would be limited only by the output saturation of the amplifier, and the resulting distortion would be quite high. In Fig. 6.1 the increasing signal turns on the zener diode feedback for the desired gain reduction. A resistor in series with the zener diode limits the gain reduction to prevent distortion from a drastic gain change.

For this sine-wave generator the frequency is determined by the feedback to the noninverting amplifier input, and the amplitude is primarily set by the feedback to the inverting input. The accuracy and stability of the waveform frequency are controlled by the characteristics of the Wien-bridge elements. A maximum frequency of operation is set by the slewing-rate limit of the operational amplifier. Waveform amplitude will be slightly greater than 1.5 times the zener diode voltage. While the exact amplitude can be varied slightly by adjusting the negative feedback, this also greatly affects distortion. Generally, the feedback resistors are chosen for low distortion, and the zener voltage is selected for the desired amplitude. In this way distortion can be reduced to approximately 0.5 percent with this circuit.

A significant reduction in distortion can be achieved by using an AGC loop that produces the ideal gain level, rather than the clamping of the above zener diode approach. Distortion is introduced with the zener clamp techniques by the abruptness of the gain change accompanying zener turn-on and by the positive feedback gain of greater than unity needed to initiate oscillation. To also avoid distortion from excessive gain, active AGC loops are used which reduce gain to the optimum level following turn-on. This action is achieved using a JFET as a voltage-controlled resistor in Fig. 6.2.[3] The JFET is well suited because its

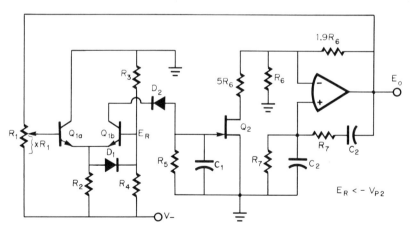

$E_o = A \sin \omega t$ $\omega = 1/2\pi R_7 C_2$ $A = E_R/x + (1 - 1/x)V_-$

Fig. 6.2 Amplitude stabilization and control are provided for a Wien-bridge oscillator by an AGC circuit formed with a differential stage comparator and an FET voltage-controlled resistor.

resistance can be varied from approximately 500 Ω to 100 MΩ in response to a gate control voltage. That voltage is derived from the oscillator output through the comparator action of the differential stage formed with Q_{1a} and Q_{1b}. Comparison is made between the attenuated oscillator signal at one comparator input and the reference voltage E_R at the other input. When the negative peaks of the oscillator signal are too small, Q_{1b} will not be turned on. No current is then supplied to maintain a control voltage on C_1, and the FET appears as a low resistance. This shunts R_6 to temporarily raise the oscillator gain and initiate oscillation or increase amplitude.

When the negative signal peaks become large enough to trigger the comparator, Q_{1b} is pulsed on to develop a control voltage on C_1. This voltage begins turning the FET off and raising its resistance. At some signal amplitude the FET resistance is that which makes the net gain around the positive feedback loop unity, and the amplitude stabilizes. The gain remains near unity throughout the oscillation cycle as the filter capacitor holds the gate bias between peaks. For this operation the FET must be the deciding factor on whether oscillation occurs; so gain without the FET must be set slightly lower than required for oscillation.

By means of the control potentiometer R_1, the oscillator amplitude can be varied from zero to the output saturation level of the amplifier. This potentiometer attenuates the signal reaching the comparator from the oscillator and biases the comparator input toward the negative supply. When adjusted so that this comparator input is connected directly to the oscillator output, the comparator triggers at an output peak level equal to the reference voltage E_R. This sets the oscillator amplitude at E_R, unless amplifier saturation imposes a lower limit. As adjustment of R_1 shifts the comparator input toward the negative supply, less oscillator swing is required to drive the comparator to its trip point; so amplitude is reduced. At some point along R_1, the dc bias alone is sufficient to trigger the comparator, and the oscillator amplitude is reduced to zero. This amplitude control is described by

$$A = \frac{E_R}{x} + \left(1 - \frac{1}{x}\right)V_- \quad \text{for } x \geq 1 - \frac{E_R}{V_-}$$

Alternatively, the amplitude can be voltage-controlled, for modulation or automatic systems, by replacing E_R with a control voltage.

Amplitude accuracy and stability are primarily determined by the AGC circuit. Gain provided by the comparator reduces the portion of the output signal required to create the FET control voltage. As a result, output amplitude is much less dependent upon thermal variations in the FET characteristic. The net gain in the AGC feedback loop is reduced by the attenuation of R_1 and is lowest for small amplitudes. But the gain remains sufficient to limit amplitude variations following warm-up to about ±10 mV. Temperature changes induce approximately a 1 mV/°C additional amplitude variation because of the drift of the comparator transistors. Since these transistors are not operating at fixed, matched currents, their differential drift is relatively high. For laboratory environment this amplitude stability permits a 100:1 amplitude range with 0.1 percent of full-scale error.

This amplitude stability is compatible with the output distortion level. Distortion results from signal-induced variations in the FET resistance and from the decay of the gate control voltage between cycles. The voltage-sensitive FET resistance varies with the signal voltage between drain and source causing oscillator gain variation. Because of the gain variation the positive and negative peaks will be of slightly different amplitudes. A similar distortion is produced by the decay in the control voltage at the FET gate between recharging intervals at the negative peaks. This effect also makes for gain variation during the signal cycle. Reduced control voltage decay is achieved with a greater holding time constant R_5C_1, but this also increases the response time for amplitude adjustment.

By means of an added amplifier, the frequency of a Wien-bridge oscillator can be controlled with one resistor. Normally, it is not convenient to vary the frequency of these oscillators, because changing any one of the frequency-determining elements upsets the critical gain of the circuit. The net gain through the negative and positive feedback paths of a Wien-bridge oscillator must be accurately maintained at unity to permit amplitude stabilization as described above. If any one element is varied to change the oscillator frequency, another must also be accurately adjusted to restore the feedback loop gain. Thus, variable frequency with the common Wien-bridge oscillator would require use of precisely matched ganged potentiometers and would generally be a poor economic choice for sine-wave generation.

Instead, the Wien-bridge oscillator can be adapted to variable-frequency operation with a control amplifier that simultaneously alters the frequency-determining Wien bridge and the feedback-loop gain.[4] One potentiometer makes both adjustments when connected with the control amplifier as in Fig. 6.3. The potentiometer R_2 serves as part of the Wien bridge and as the gain control for the amplifier formed with A_1. Since R_2 is part of the bridge, it influences the oscillator frequency, as expressed by

$$f = \frac{1}{2\pi C \sqrt{R_1 R_2}}$$

As R_2 is decreased to increase frequency, it reduces the signal fed back as e_B. However, it simultaneously increases the gain by which e_B is amplified to maintain a net unity gain around the feedback loop. The output

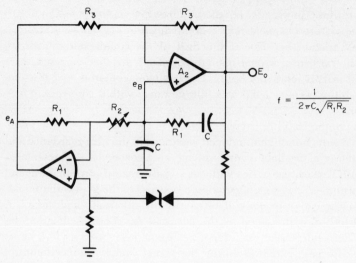

Fig. 6.3 A single-resistor frequency control for a Wien-bridge oscillator is made possible with a gain control amplifier that simultaneously varies the Wien-bridge and the feedback-loop gain.

voltage is attenuated by the bridge to develop e_B, and at the oscillation frequency that attenuation equals

$$\frac{e_B}{E_o} = \frac{R_2}{R_1 + 2R_2}$$

Both amplifiers amplify e_B to develop the output voltage as expressed by

$$E_o = 2e_B - e_A = \left(2 + \frac{R_1}{R_2}\right)e_B = \frac{R_1 + 2R_2}{R_2}e_B$$

Combining the last two equations reveals that the net loop gain of the oscillator is unity, independent of the value of R_2. Thus, the loop gain and, therefore, amplitude are not affected by the frequency control.

In practice, the loop gain is sensitive to oscillation frequency because of stray capacitances and amplifier bandwidth limitations. These factors limit the practical range of frequency adjustment for constant output amplitude. With the simple zener diode AGC feedback shown, relatively constant amplitude is maintained for a 10:1 frequency range. With more refined AGC the usable frequency range can be extended, but it will still be far less than the range over which oscillation can be sustained by the circuit.

6.2 Square- and Triangle-Wave Generators

Among the most common signals generated with operational amplifiers are square waves and triangle waves, because their generation relies on two basic operational amplifier functions, integration and voltage comparison. A triangle wave results from integration of a square wave, which in turn is generated by voltage comparison of the triangle wave against a reference switched by hysteresis. Thus integration and comparison in a common feedback loop generate the two desired signals. Utilizing this approach are the generators of this section, which feature improved symmetry, unipolar or trilevel square waves, and swept-frequency outputs.

Both the integration and comparison functions can be performed with one operational amplifier in generating square and triangle waves, but usually at the expense of triangle-wave linearity. Improved linearity is achieved through the use of FET current sources for feedback,[1] but the signal symmetry is degraded by the mismatch between the two FETs. This remaining error can be removed by making use of the symmetrical nature of many JFETs. For FETs the drain and source are essentially interchangeable without significant effect on drain current magnitude. It is this feature that makes possible the use of one FET for both polarities of capacitor charging current in the square- and triangle-wave generator of Fig. 6.4.

By means of diodes, the FET gate is switched between drain and source circuits to switch the polarity of current supplied to the capacitor. When the amplifier output is in its negative state, D_1 forward-biases to connect the gate to the source through a bias resistor. This connects the FET in its normal current-source configuration, and it supplies a positive current i_c to the capacitor. By appropriate choice of the bias resistor R_S, the current i_c is made equal to the zero temperature coefficient current of the FET for

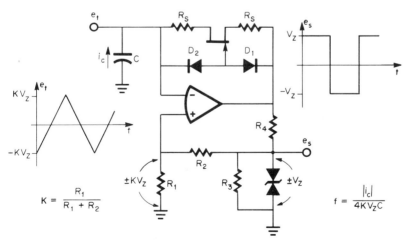

Fig. 6.4 A single-amplifier triangle- and square-wave generator with improved waveform symmetry results from diode-switching one JFET to supply both polarities of capacitor charging current.

temperature stability. When the capacitor voltage reaches the threshold $-KV_z$ set by the hysteresis feedback, the amplifier output switches positive, reverse-biasing D_1 and forward-biasing D_2. In this state the FET operates in its inverted mode to supply the opposite polarity current to the capacitor. This current charges the capacitor from a voltage of $-KV_z$ to $+KV_z$, where the hysteresis feedback establishes the second threshold. At that voltage, the amplifier output switches back to its negative state. As the capacitor charging and discharging repeats, output waveforms are generated that are triangular at the capacitor and square wave at the amplifier output with a frequency of

$$f = \frac{|i_c|}{4KV_zC}$$

where

$$K = \frac{R_1}{R_1 + R_2}$$

These waveforms are symmetrical with respect to time for the common symmetrical characteristic of JFETs.

With capacitor charging linearity and symmetry controlled by the FET, the resulting waveform characteristics are dependent upon the FET output resistance and the voltage match of the dual zener diode. As the capacitor voltage rises, the voltage across the FET decreases, and capacitor charging current varies in accordance with the FET output resistance. Such current variation produces nonlinearity in the triangle waveform slope, but this is reduced from the common circuit by the ratio of this output resistance to that of the resistor normally used in place of the FET cur-

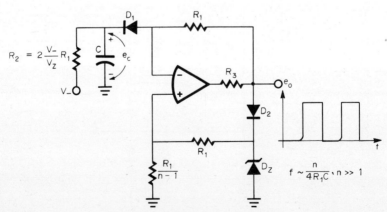

Fig. 6.5 A zero-based square wave is generated by the common single-amplifier oscillator when the circuit modes are altered by gating diodes.

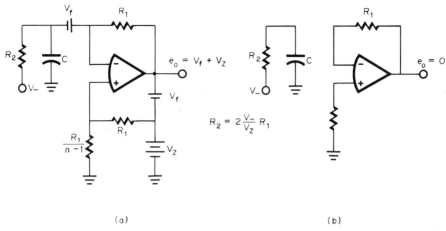

FIG. 6.6 Through diode gating action, the circuit of Fig. 6.5 switches from a common oscillator connection to a zero output configuration for generation of a zero-based square wave.

rent source. Unequal voltage magnitudes for the two polarities of the dual zener voltage result in unequal magnitudes for the two circuit switching points. Directly related to this difference is the time difference of the two circuit states that determines waveform dissymmetry with respect to time. Also for timing accuracy, the operational amplifier input impedance must remain high under input overload. Added to the circuit of Fig. 6.4 are R_4 and R_3, which limit amplifier output current and discharge zener capacitance, respectively.

With the common single-amplifier square-wave generator, a bipolar square wave is generated. While this is often desirable, it is not compatible with many circuits, such as digital logic. For such applications, the basic oscillator can be modified for a unipolar square-wave output with a precisely set zero base, although with less precise frequency. That modified circuit is illustrated in Fig. 6.5, which adds signal gating diodes D_1 and D_2. When the amplifier output is positive, both diodes are forward-biased, resulting in a circuit connection like that of the basic single-amplifier square-wave generator.[1] With the diodes forward-biased, the circuit is represented by Fig. 6.6a, and the amplifier output voltage charges the capacitor until the switching threshold set by the positive feedback is reached.

At that threshold of V_Z/n, the amplifier output swings negative, reverse-biasing the diodes to result in the circuit mode of Fig. 6.6b. Now the operational amplifier is connected as a voltage follower with zero input voltage, and the output resides at zero volts, rather than the negative level of the basic circuit. While in this mode, the capacitor is discharged through R_2 until its voltage reaches a level that again forward-biases D_1

of Fig. 6.5 to drive the amplifier output positive. This state-switching oscillation continues, developing a zero-based square-wave output with a frequency on the order of

$$f \approx \frac{n}{4R_1C} \qquad n \gg 1$$

Note that the oscillation frequency is approximated by this expression only when the positive feedback ratio $1/n$ is small. Small positive feedback, however, results in a rounding of the signal at the beginning of the transition from the zero voltage state to the positive stage. This rounding can be reduced by increasing positive feedback, for which the frequency-defining expression becomes more complex.

Diode gating of the basic operational amplifier oscillator can be extended from that of Fig. 6.6 to develop a three-state output, as in Fig. 6.7. In this case separate timing capacitors are used for the positive and negative output states, and the capacitors are diode-gated to allow an intermediate zero voltage state. At turn-on, the capacitor voltages are zero; so neither diode is forward-biased. This leaves the amplifier connected with unity negative feedback and less than unity positive feedback. Essentially, this connection is that of a voltage follower with positive feedback, and the output voltage will follow V_T. But the positive feedback establishing V_T sets this voltage at some fraction of the output voltage; so there are two constraints defining the output level in this circuit state:

$$e_o = V_T$$
$$V_T = \frac{e_o}{n}$$

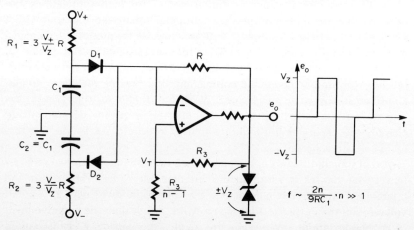

Fig. 6.7 To generate a trilevel output, a common oscillator configuration is modified to include two timing capacitors that are diode-gated to the amplifier input.

The only condition satisfying these two constraints is $e_o = 0$ for $n > 1$; so the output remains at zero voltage when the two diodes are reverse-biased.

A zero voltage output state continues until one of the capacitors is charged to a voltage that forward-biases a diode. If the power-supply voltage magnitudes are equal and the R_2C_2 time constant is less than R_1C_1, C_2 will reach the turn-on voltage of D_2 before C_1 reaches its corresponding level. Then, the amplifier output is driven positive, where it supplies hysteresis feedback to set the next switching threshold at $V_T = V_Z/n$. That threshold is reached when C_2 has been charged to that voltage by means of current supplied through D_2. At that threshold, the amplifier output swings negative, reducing V_T and reverse-biasing D_2. This negative swing would stop at zero except that C_1 has been, meanwhile, charging and has a voltage large enough to forward-bias D_1. As a result, the amplifier output continues its swing to its negative voltage state, where it supplies a hysteresis feedback voltage $V_T = -V_Z/n$. That state continues until C_1 has been discharged to this new hysteresis level through D_1, at which time the output swings positive again. However, this swing stops at the zero voltage level because C_2 has not yet recharged to a voltage great enough to forward-bias D_2. To ensure that the intermediate zero voltage output state remains in the circuit equilibrium operation, the capacitor charging and discharging are ratioed through resistor selection. When a given diode is forward-biased, it discharges the capacitor through the operational amplifier feedback resistor R from a voltage equal to V_Z. That discharge lasts the length of one output state. Recovery from this discharge is controlled by resistors R_1 and R_2. This recovery is made to last a time equal to the duration of two output states by relating the resistances of R_1 and R_2 to the feedback resistor R and charging voltages as shown. Such ratioing makes the time durations of the three output states equal. Other resistance ratios can be used to produce unequal state durations, but the capacitor voltage recovery times must always be made longer than the discharge time to ensure presence of the zero voltage state.

A square- and triangle-wave generator having a swept-frequency output can be formed by combining a ramp generator with a voltage-controlled oscillator. As the ramp voltage increases with time, it drives the voltage-controlled oscillator for linearly increasing frequency as in Fig. 6.8. Forming the ramp generator are A_1 and A_2, which perform as an integrator and a comparator in a common feedback loop.[1] Current supplied through R_3 results in an integrator output voltage that increases linearly with time until that voltage reaches the comparator threshold. That threshold is controlled by the comparator positive feedback and the bias from the potentiometer R_N to be $V_M + V_{Z1}$. When this threshold is reached, the comparator switches to its positive output state to supply an integrator discharge current through the signal diode and R_2. For rapid reset R_2 is made

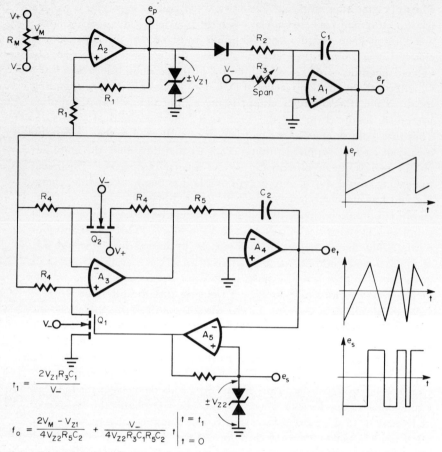

Fig. 6.8 Swept-frequency square and triangle waves are produced by a ramp generator driving a voltage-controlled oscillator.

small compared with R_3. Reset continues until the second comparator trip point is reached at $V_M - V_{Z1}$, where the circuit switches back to generate another ramp.

To generate a swept-frequency output, this ramp drives the voltage-controlled oscillator formed with A_3, A_4, and A_5.[1] Together A_4 and A_5 would form a square- and triangle-wave generator with operation much like that described for the ramp generator above. When the integrator output voltage reaches the trip points of comparator A_5, the latter switches to reverse the direction of integration. In this circuit it produces the integration reversal by switching the polarity of the gain provided by A_3. When switch Q_1 is on, the noninverting input of A_3 is grounded, and this amplifier acts as an inverter to supply the integrator with the ramp voltage in inverted form. In this mode, Q_2 serves as a resistor to compensate for the error introduced by the ON resistance of Q_1.

By turning Q_1 off, the role of A_3 is switched from inverter to voltage follower for the desired gain polarity reversal. With Q_1 off the ramp signal is connected to the noninverting input of A_3 through a resistor. Since no current is drawn through that resistor, the ramp signal appears direct at the noninverting input and at the inverting input because of feedback. Now both ends of the resistor connected to the inverting input have the same signal voltage; so the resistor conducts no signal current from the feedback resistor. No signal is then developed on the R_4 feedback resistor, and the output voltage of A_3 equals the ramp voltage.

Since the ramp generator output from A_1 controls the magnitudes of both positive and negative integrator inputs, it controls both integration rates. Increasing ramp voltage results in greater integration rates and shorter times spent between comparator trip points for higher frequency oscillation. That frequency varies linearly with time and is expressed by

$$f_o = \frac{2V_M - V_{Z1}}{4V_{Z2}R_5C_2} + \frac{V_-}{4V_{Z2}R_3C_1R_5C_2} t \Big|_{t=0}^{t=t_1}$$

where

$$t_1 = \frac{2V_{Z1}R_3C_1}{V_-}$$

Note from this expression that the lower frequency limit can be controlled by V_M, which by itself sets the midpoint of the frequency sweep. To control the span of the frequency sweep, R_3 is adjusted.

6.3 Ramp and Pulse Generators

A variety of free-running reference controlled ramp and pulse generators are formed with operational amplifiers.[1] Added to these are the *triggered ramp generator* and the *digitally controlled ramp generator* of this section. Conceptually, a triggered ramp generator can be formed with an integrator that is driven by a constant voltage for a controlled time period in response to a trigger signal and is then reset. An applied voltage for a controlled time is readily achieved with a monostable multivibrator, or *one-shot*. The integrator reset can be provided by a feedback amplifier, which also eliminates equilibrium state drift. Both the timing and reset functions can be performed with one amplifier that switches roles in the triggered ramp generator of Fig. 6.9.

In this circuit A_2 operates as an integrator controlled by the one-shot feedback amplifier A_1. At equilibrium, capacitor C_2 has discharged so that D_1 is zero-biased, leaving A_1 in a noninverting amplifier configuration.

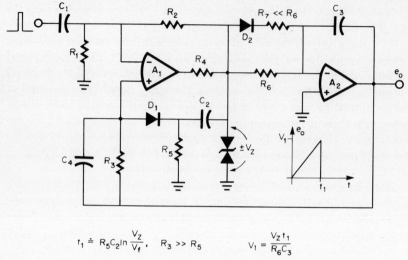

$$t_1 \doteq R_5 C_2 \ln \frac{V_z}{V_f}, \quad R_3 \gg R_5 \quad V_1 = \frac{V_z t_1}{R_6 C_3}$$

Fig. 6.9 A triggered ramp generator is formed with an integrator that receives a constant voltage input from a one-shot upon application of a trigger pulse.

This amplifier senses the integrator output voltage through R_3 and forces that output to remain at essentially zero voltage. Since the integrator and feedback amplifier are in a common feedback loop, care must be taken to preserve feedback stability, generally through the use of large integrator time constants. Lower feedback amplifier closed-loop gain also helps in this effort, because bandwidth is then greater; but high gain is desirable for faster switching in the one-shot mode.

When a positive trigger voltage is applied to A_1, it switches to its one-shot operation. The output of A_1 then swings negative, coupling a positive feedback signal through C_2 that forward-biases D_1 and holds the amplifier output in its negative state. This applies a voltage of $-V_z$ to the R_6 integrator input resistor to develop a linearly increasing integrator output voltage. Integration continues until the voltage on C_2 decays to turn D_1 off again and disconnect the positive feedback. This occurs after a time

$$t_1 \doteq R_5 C_2 \ln \frac{V_z}{V_f} \quad \text{for } R_3 \gg R_5$$

During this time, the ramp reaches a peak voltage of

$$V_1 = \frac{V_z t_1}{R_6 C_3}$$

At the time t_1, D_1 reverse-biases, returning A_1 to its noninverting amplifier mode, where it senses the integrator output. Initially, this output resides at the ramp peak voltage. So A_1 swings positive to forward-bias D_2 and discharge the integrator capacitor through R_7. This resets the integrator to zero until the next trigger pulse is applied.

Digital control over the frequency of a ramp and pulse generator can be achieved by using a digital-to-analog converter to control the generator integration rate. Such an approach is illustrated in Fig. 6.10, where a digital-to-analog converter is combined with a common ramp and pulse generator circuit.[1] The digital input word applied determines the voltage level presented by the converter to the integrator formed with A_1. Because the integrator input voltage controls the output rate of change, the digital-to-analog converter controls the time required for ramp voltage e_r to swing from one comparator trip point to the other. As long as this time is much greater than that required for reset, the ramp duration represents a signal period related to a digital input. The associated signal frequency is directly proportional to the digital input

$$f = \frac{R_4 M}{2R_1 R_3 C V_Z} \Sigma_1^n \frac{1}{2^i} b_i$$

where M is the gain constant of the digital-to-analog converter.

For the above expression to be accurate, the reset time must be negligible, and this is controlled by the choice of R_1 and R_2. Reset occurs when the ramp output reaches the negative comparator trip point and the comparator output switches negative. This forward-biases D_1 to conduct

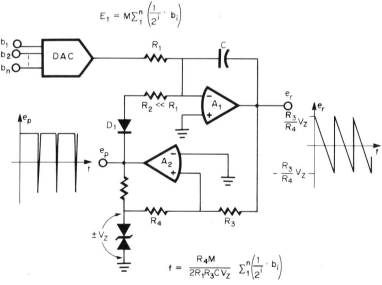

Fig. 6.10 For digital control of the frequency of a ramp and pulse generator, a digital-to-analog converter is used to control the integration rate of the generator.

an integrator resetting current through R_2, which must be much smaller than R_1 for negligible reset time. For reset by this negative comparator output signal, the digital-to-analog converter output must be positive. If, instead, the converter output is negative, D_1 should be reversed for reset from the positive comparator output. In the latter case, the resulting ramp and pulse waveforms will be inverted from those shown.

6.4 Staircase Generators

A more specialized form of a ramp signal is a *staircase* waveform, which increases in discrete steps rather than linearly. As such, the staircase signal supplies step increases in voltage level during consecutive time periods, as is useful for sequential control and multiple level testing. Described in this section are staircase signal generators of varying complexity for different accuracy levels and step interval lengths.

With one operational amplifier an elementary staircase generator can be formed for applications where a square wave or pulse train is available to drive the circuit. Shown in Fig. 6.11, this generator resembles the basic half-wave rectifier except that capacitive input and feedback elements are used. To convert a rectangular wave input signal into a staircase, this circuit differentiates the input signal, rectifies the result, and then reintegrates it. Differentiation is performed by C_1, which converts the rectangular waveform into pulses of current. To limit these current pulses for prevention of amplifier overload, resistor R is added. While the resistor does alter the wave shape of the current pulses, it does not affect the charge transferred by those pulses so long as the time constant RC_1 is small in comparison to the signal period.

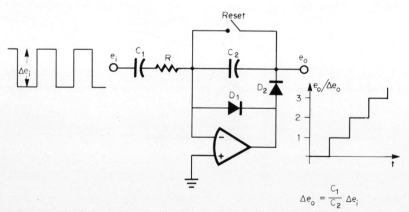

Fig. 6.11 An elementary staircase generator differentiates, rectifies, and reintegrates the signal from an input rectangular waveform.

Signal Generators

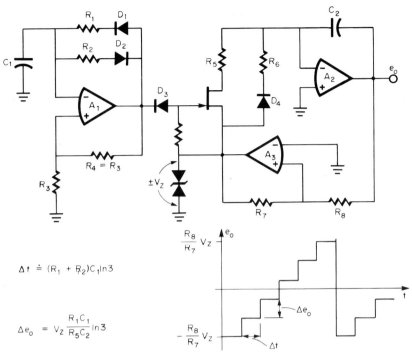

Fig. 6.12 A complete staircase generator is formed with a step-controlling pulse generator, an integrator, and a reset comparator.

Only one polarity of the current pulses passes through C_2 to be integrated. That is the polarity that forward-biases D_2 and reverse-biases D_1 to connect C_2 in the signal path. Opposite polarity current is conducted by D_1 direct to the amplifier output, bypassing C_2. Unipolar current pulses result in step increases in the voltage on C_2, as desired. Step increments are determined by the charge transferred through C_1 to C_2, and

$$\Delta e_o = \frac{C_1}{C_2} \Delta e_i$$

While this is a relatively simple means of generating a staircase waveform, performance is limited by input signal amplitude error and by output loading. Current for output loads is supplied by C_2 when D_2 is not conducting, and this produces output voltage droop. In addition, the circuit requires separate provision for supply of the input rectangular wave and for circuit reset. A more complete staircase generator is formed with three operational amplifiers as in Fig. 6.12. To generate a staircase waveform, these amplifiers function as a pulse generator, an integrator, and a comparator.

The pulse generator,[1] formed with A_1, produces step increments by turning on an FET switch that connects an input resistor of integrator A_2 to a reference voltage of $-V_Z$. That reference voltage is negative because of the state of comparator A_3 during the staircase rise. Pulse duration and period are chosen to set the step rise time and duration which will be

$$t_r \doteq R_1 C_1 \ln 3$$
$$\Delta t \doteq (R_1 + R_2) C_1 \ln 3$$

During the time t_r, integrator input resistor R_5 is connected to a voltage of $-V_Z$ by the FET switch. This produces an output rise of

$$\Delta e_o = V_Z \frac{R_1 C_1}{R_5 C_2} \ln 3$$

Repeated steps in voltage are produced at intervals of Δt until the output voltage reaches $R_8 V_Z / R_7$, where it trips comparator A_3. Then the comparator output switches to its positive state, forward-biasing D_4 to reset the integrator output to a voltage of $-R_8 V_Z / R_7$. At that voltage, the comparator switches back to its negative output state, and the staircase begins again.

Another staircase generator circuit is available that has both positive- and negative-going staircase outputs as shown in Fig. 6.13. This circuit

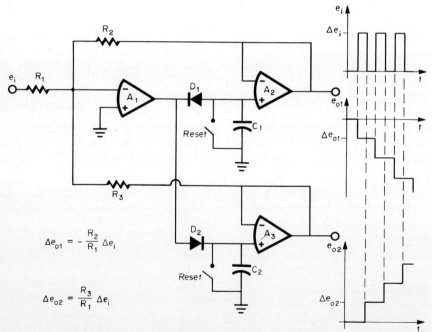

Fig. 6.13 Positive- and negative-going staircase waveforms are generated by a circuit that performs as both a positive and a negative peak detector.

requires a zero-based rectangular wave input, and step amplitude is related to that of the rectangular wave. Peak detector circuitry is the basis for this staircase generator. An inverting, positive peak detector is formed by A_1 and A_2, and A_3 forms an inverting, negative peak detector with A_1. Signals from the two output amplifiers are feedback-summed at the input to result in staircase generation.

The first positive input swing is detected by A_1 and A_2, so that a negative voltage is established on C_1. When the input signal returns to zero, the negative voltage on C_1 drives the input of A_1 through R_2 and A_2. In this mode, C_1 is supplying a negative input to the inverting, negative peak detector formed with A_1 and A_3. Thus, a positive voltage is established on C_2 which biases R_3 through A_3 to return the input of A_1 to zero. When the second positive input swing occurs, it again raises the input voltage of A_1, so a more negative voltage is developed on C_1 to restore zero voltage at the summing junction. This greater negative voltage on C_1 must be counterbalanced by a greater positive voltage on C_2 when the input signal returns to zero. Thus, the second step is generated at both outputs. Further input pulses continue to produce output steps in the same way as controlled by the summing feedback to the input A_1.

Circuit requirements and limitations are related to those of peak detectors as described in Sec. 3.2. Critical to step amplitude accuracy is phase compensation selection for zero overshoot. As with any peak detector circuit, overshoot can greatly overcharge the storage capacitors; so amplifier damping should be greater than normal. Step droop and rise-time limitations, as well as means to reduce them, are described earlier for peak detectors.

The above staircase generators accumulate voltage on a capacitor that receives an increment of charge at regular time intervals. This approach is less suitable for long time intervals, such as those in sequential heater control, because parasitic current drains on the capacitor induce voltage droop. One means of avoiding droop error in staircase generation is to use a digital-to-analog converter driven by a clocked, digital counter. While this ensures high degrees of voltage step and timing accuracies, it is less economical than often desired.

A more economical approach is illustrated in Fig. 6.14. With this circuit, precise control is maintained over the voltage levels of the steps; however, step timing accuracy is somewhat compromised for the sake of circuit simplicity. Basically, the circuit consists of an inverting amplifier with a number of summing resistors that are sequentially connected to a reference voltage. As each summing resistor is connected, the circuit output voltage rises by a precise voltage increment.

The connections are made by means of MOSFET switches driven from a simple RC timing element. Following reset of the timing capacitor,

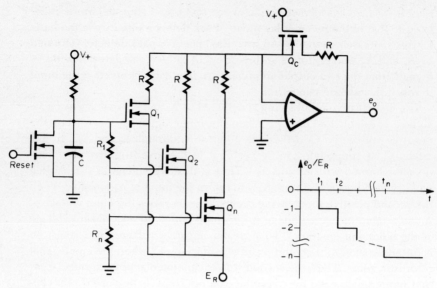

Fig. 6.14 To generate a staircase with long duration steps, the gain applied to a reference voltage is sequentially switched as a timing capacitor charges.

its voltage is too low to turn on any of the FETs; so the amplifier output remains at zero volts. When the capacitor voltage reaches a level equal to the reference voltage E_R plus the threshold voltage of Q_1, that FET turns on, connecting E_R to one summing resistor. From this connection an output voltage of $-E_R$ is developed, if the ON resistance effect of Q_1 has been well compensated for by that of Q_C. The compensation FET Q_C adds a resistance to the feedback resistor that approximately equals that resistance added to the summing resistors by Q_1 through Q_n. As long as the feedback and summing resistors are all equal in value, matched ON resistances will produce no error. Because the ON resistances cannot be exactly matched, their effects should be further minimized by making the resistances of the summing and feedback resistors large compared with the ON resistance.

As the capacitor voltage increases further, its attenuated output to the gate of Q_2 becomes large enough to turn on that FET. Then, two summing resistors are connected to E_R; so the circuit output voltage becomes $-2E_R$. Continuing capacitor voltage increase sequentially turns on each FET to step the output voltage in increments of $-E_R$. While these voltage increments are precise, the time increments are sensitive to the tolerance and thermal variations of the MOSFET threshold voltages. Timing accuracy also depends upon the choice of resistors R_1 through R_n to match the RC charging wave shape of the simple timing elements.

6.5 Timing Circuits

For electronic control of event time duration, monostable multivibrators, or one-shots, are widely used. These timing circuits can be formed with operational amplifiers for high precision and circuit simplicity.[1] In this section, monostable multivibrators are illustrated for moderate- or high-precision requirements and for level, rather than pulse, triggering.

In simplest form, a pulse-triggered one-shot is formed with capacitive positive feedback as in Fig. 6.15. At equilibrium, no feedback current flows in the capacitor, and the amplifier output is held in its negative state by the bias of R_1 and R_2. This bias fixes the triggering threshold voltage at

$$V_T = \frac{R_2 V_-}{R_1 + R_2}$$

Note that this voltage cannot be set greater in magnitude than the forward voltage drop of D_1, because that diode would clamp the voltage at this point. If a larger threshold voltage is needed, the diode can be removed at the expense of increased recovery time.

When a negative trigger pulse drives the inverting amplifier input below the level of V_T, the amplifier output swings positive. This couples a positive feedback signal through C to latch the circuit in the positive state. That state continues until C has discharged to the point at which the noninverting input reaches zero voltage. Then, the amplifier output switches back to its negative state, completing a pulse of duration

$$t_1 = (R_1 \parallel R_2) C \ln \frac{2V_Z}{V_T}$$

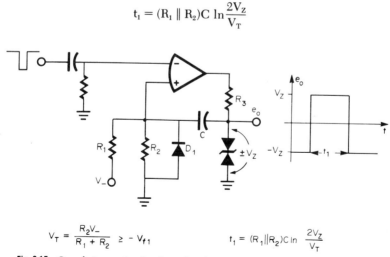

$$V_T = \frac{R_2 V_-}{R_1 + R_2} \geq -V_{f1} \qquad t_1 = (R_1 \parallel R_2) C \ln \frac{2V_Z}{V_T}$$

Fig. 6.15 One-shot operation is achieved with an operational amplifier having capacitive positive feedback.

170 Designing with Operational Amplifiers

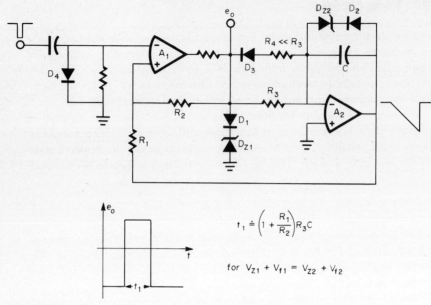

Fig. 6.16 Monostable multivibrator performance with a square- and triangle-wave generator results when the integrator output is clamped away from one trip point of the comparator.

At the end of the time pulse, C begins recharging to its equilibrium. This recharging time, or recovery time, must occur before the circuit is again triggered, or timing will be in error. To shorten this recovery time, diode D_1 shunts R_2. This recovery also requires high current from the operational amplifier output, so limiting resistor R_3 must be chosen considering recovery time as well. Timing accuracy is also dependent upon the current drawn by the operational amplifier input, so the amplifier should be chosen for high input resistance under input overload.

Immunity to trigger pulses that occur during a timing pulse and higher accuracy result with the two-amplifier monostable multivibrator of Fig. 6.16. This circuit is essentially a square- and triangle-wave generator[1] whose integrator output is clamped. Without the clamping of D_{Z2}, the integrator output would swing back and forth between the comparator trip points in response to the switching comparator output. By clamping the integrator swing in one direction, one prevents it from reaching one of the comparator trip points; so circuit oscillation is inhibited.

A negative trigger pulse reaching the inverting comparator input shifts the trip point to within the range of the clamped integrator output voltage, and the comparator then switches to its positive output state. In this state hysteresis feedback through R_1 and R_2 shifts the comparator trip point to a positive level where it is immune to negative trigger pulses. Positive trigger pulses, which could cause triggering during this timing interval, are kept from reaching A_1 by the clamping of D_4. Thus, the output of A_1 remains positive until the integrator output reaches that positive trip point. The duration of this positive state is

$$t_1 \doteq \left(1 + \frac{R_1}{R_2}\right) R_3 C \quad \text{for } V_{Z1} + V_{f1} = V_{Z2} + V_{f2}$$

As expressed, timing is dependent only upon passive component accuracy if the diode voltage match indicated is maintained. Shortened recovery time is achieved by adding D_3 and R_4 for faster integration rate when the comparator output is negative. With the signal available from the integrator output this circuit is also useful as a triggered ramp generator.

In some electronic control functions it is desirable to have a timing pulse triggered when a monitored signal reaches a certain level. This can be realized in a straightforward manner by using a comparator to generate a trigger pulse for a one-shot when the reference level is reached. However, both the comparator and one-shot functions can be performed with the same operational amplifier using the circuit of Fig. 6.17. This circuit switches from a comparator mode to a timer mode when the comparator trip point is reached.

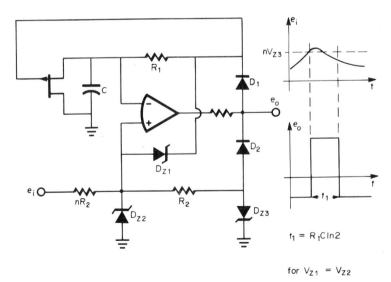

Fig. 6.17 A level-triggered one-shot results with an operational amplifier having both comparator and timing feedback.

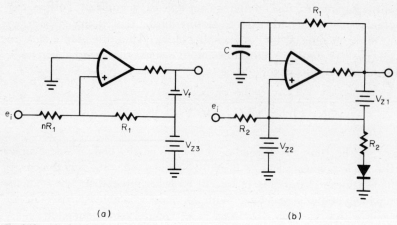

Fig. 6.18 When the input signal to the circuit of Fig. 6.17 reaches a threshold level, the circuit switches from (a) a comparator connection to (b) a timing circuit.

When the input signal e_i is below the trip point, the amplifier output is negative. The negative output voltage forward-biases D_2 to connect the comparator hysteresis feedback of R_2 and nR_2, and reverse-biases D_1 to disconnect the timing feedback. Also, the negative voltage lets the FET switch turn on to ground the amplifier inverting input, and the circuit is represented by Fig. 6.18a. In this mode the circuit appears as a comparator with a trip point at nV_{Z3}.

When the input signal reaches this trip point, the amplifier output swings positive, resulting in a timing circuit connection. The positive output reverse-biases D_2 to disconnect the comparator feedback, forward-biases D_1 to connect the timing feedback, and turns off the FET switch. Representing this mode is Fig. 6.18b. Initially, the capacitor voltage is zero, since it was shorted in the previous state; so the amplifier inverting input voltage is below that at the noninverting input. This voltage difference holds the amplifier output positive until the capacitor charges through R_1 to make the inverting input voltage slightly larger V_{Z2}. Then the amplifier output switches back to its negative state. The result is a positive state duration of

$$t_1 = R_1 C \ln 2 \quad \text{for } V_{Z1} = V_{Z2}$$

If the input signal remains above the comparator trip level, the one-shot action will repeat. For this reason, the circuit is primarily useful where the one-shot output is used to perform a function that in turn lowers e_i, such as in feedback control.

REFERENCES

1. J. Graeme, *Applications of Operational Amplifiers: Third-Generation Techniques*, McGraw-Hill Book Company, New York, 1973.
2. G. Tobey, J. Graeme, and L. Huelsman, *Operational Amplifiers: Design and Applications*, McGraw-Hill Book Company, New York, 1971.
3. J. Graeme, AGC Provides 0.1% Amplitude Stability for Wien-Bridge Oscillator, *Electron. Des.*, May 24, 1975.
4. P. Brakow, FET Op Amp Adds New Twist to an Old Circuit, *EDN*, June 5, 1974.

7
COMPUTING CIRCUITS

Some of the earliest and most widespread applications of operational amplifiers involved analog computation. With operational amplifiers, functions combining algebraic, differential, and trigonometric terms are implemented. Described in this chapter are techniques for addition, subtraction, integration, differentiation, multiplication, division, power generation, root taking, and trigonometric response generation. Expanded forms of summing amplifiers, difference amplifiers, integrators, and differentiators are illustrated that extend the computing ability of such circuits. Logarithmic techniques are applied for simplified multiplication, division, power generation, and root taking. With the power and root functions, trigonometric approximations through power-series expansion are presented for sine, cosine, and arctangent functions. Concluding the chapter are circuit implementations that simplify vector magnitude and rms computation.

7.1 Adders and Subtractors

Perhaps the simplest computing functions performed with operational amplifiers are addition and subtraction. By virtue of the high open-loop gain of these amplifiers, very little signal is required between the amplifier inputs to develop the output voltage dictated by feedback con-

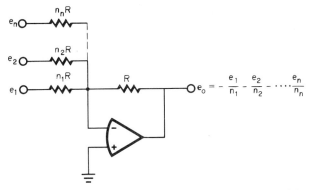

Fig. 7.1 The generalized summing inverter adds input signals while providing weighted gains.

straints. This characteristic makes possible signal summation and subtraction through the simple connection of summing or differencing resistors to the amplifier inputs. With appropriate choice of resistor values, summation with weighted gains is provided, and subtraction can be extended to an arbitrary number of input signals.

For signal addition with weighted gains, the basic inverting amplifier is connected with inversely weighted summing resistors as in Fig. 7.1.[1] As indicated, the various input signals are inverted and amplified by the ratio of the feedback resistor to the individual summing resistors. This approach can be extended to an arbitrary number of input signals, depending upon limitations imposed by the associated increased gain, offset, and bandwidth errors. Each of these error sources becomes proportionally more significant with the reduction in feedback factor[1] accompanying the addition of summing resistors. The feedback factor equals the voltage divider ratio associated with the feedback resistor and the parallel combination of the summing resistors. As summing resistors are added and feedback factor is reduced, the feedback-loop gain is decreased, and the circuit gain error increases in direct proportion. Similarly, the gain by which the input offset voltage is amplified increases proportionally with the feedback factor reduction. Further, the closed-loop bandwidth follows the same proportionality unless the amplifier phase compensation is reduced with decreases in feedback factor, as described by broadbanding techniques.[2]

Summation of signals can be combined with subtraction by extending the difference amplifier[1] connection as in Fig. 7.2. As above, multiple input resistors are connected for the desired signal combinations. However, determination of gain-setting resistors is complicated by the fact that the resistors R_{i-}, which set the negative gains, also affect positive gains. To permit arbitrary selection of both positive and negative gain

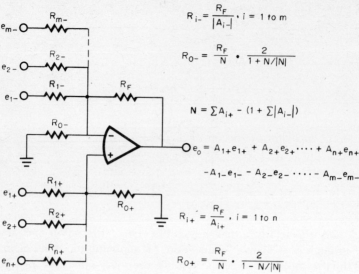

Fig. 7.2 Many signals can be added and subtracted by extending the difference amplifier to summation at both inputs.

levels, the resistor shown as R_{0-} or the one shown as R_{0+} is generally required.[3] Addition of R_{0+} results in further attenuation of the signals e_{i+} connected for positive gains. An R_{0-} resistor increases the gains to e_{i+} signals. Thus, addition of R_{0+} or R_{0-} can counteract the too-large or too-small positive gains that typically result for a given set of negative gains.

Resistor selection for the generalized adder/subtractor of Fig. 7.2 begins in a manner like that of the basic summer. For a given value of feedback resistor R_F, summing resistors R_{i-} and R_{i+} are selected for negative and positive gains A_{i-} and A_{i+} by

$$R_{i-} = \frac{R_F}{|A_{i-}|} \qquad i = 1 \text{ to } m$$

and

$$R_{i+} = \frac{R_F}{A_{i+}} \qquad i = 1 \text{ to } n$$

Then, the required positive gain compensation resistor is determined from the expressions

$$R_{0-} = \frac{R_F}{N} \frac{2}{1 + N/|N|}$$

and

$$R_{O+} = \frac{R_F}{N} \frac{2}{1 - N/|N|}$$

where

$$N = \Sigma A_{i+} - (1 + \Sigma |A_{i-}|)$$

Resulting from these equations will be an infinite value for either or both R_{O-} and R_{O+}, indicating deletion of the resistor or resistors. Neither resistor is required when the sum of the positive gains equals 1 plus the sum of the negative gain magnitudes. Then the function N defined above is zero, and both the R_{O-} and R_{O+} expressions become infinite.

As before, the number of signals that can be summed and subtracted is limited only by increasing circuit errors. Addition of summing resistors to the inverting input reduces the feedback factor, resulting in increased gain, offset, and bandwidth errors as described for the preceding circuit. Increasing the number of summing resistors to the noninverting input creates further attenuation of input signals transferred to this amplifier output. The smaller the resulting signals, the more sensitive they are to amplifier input gain and offset errors.

7.2 Integrators

One of the simplest computation functions performed with operational amplifiers is integration.[2] Variations on the basic integrator extend its computation ability to include addition, subtraction, noninverting response, and extremely large or small time constants. Such integrator variations are illustrated below.

7.2.1 Summing and noninverting integrators

To combine addition and subtraction with integration, the summing and differencing techniques of adders and subtractors are applied to integrators. Positive gain integrator operation is provided by several noninverting amplifier configurations, as will be described. For addition combined with integration, summing resistors are added at the input of the integrator as in Fig. 7.3. These resistors can be weighted in value to provide differing gains for the various input signals.

As with the summing inverter of the previous section, the number of signals that can be summed in this manner is limited only by the circuit errors that can be accepted. Addition of summing resistors lowers the feedback factor, resulting in increased gain and offset errors. Offset error is one of the most serious error limitations of integrators, because

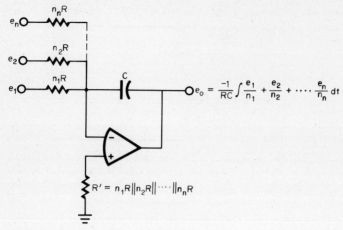

Fig. 7.3 Addition as well as integration is performed by an integrator with summing resistors.

that error is cumulative. Integration of the amplifier dc input errors results in increasing error with integration time as expressed for Fig. 7.3 by

$$\epsilon = \frac{1}{R_{eq}C} \int V_{os} \, dt + \frac{1}{C} \int I_{os} \, dt + V_{os}$$

where

$$R_{eq} = n_1 R \parallel n_2 R \parallel \cdots \parallel n_n R$$

As seen from this expression, the integrator gain by which the offset voltage is amplified is increased by addition of summing resistors since they reduce R_{eq}.

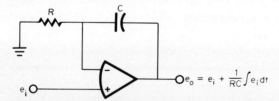

Fig. 7.4 Signal connection to the noninverting input of an integrator amplifier results in noninverting integrator operation and addition of the signal to its integral.

Where it is desired to add a signal to its integral, a simple integrator variation is available. In this case, the input signal is connected to the noninverting amplifier input rather than to the input resistor as in Fig. 7.4. This connection also provides noninverting integration and the high input impedance of noninverting operational amplifier connections. Performance is otherwise that of a basic integrator with an output offset error of

$$\epsilon = \frac{1}{RC} \int V_{os} \, dt + \frac{1}{C} \int I_B \, dt + V_{os}$$

Further signal summation can be performed by this circuit through the connection of signal summing resistors to the inverting amplifier input, as before in Fig. 7.3.

Subtraction can also be combined with integration using an integrator configuration that parallels the difference amplifier connection. With symmetrical resistor/capacitor networks connected, as in Fig. 7.5, a differential integrator is formed. This circuit takes the difference between two signals and integrates the result, as expressed in the accompanying equation. With this operation the output of a floating source can be integrated, and common-mode signals can be rejected. Common-mode rejection (CMR) is determined by both the amplifier and the integrating components. A very accurate match between like components is required to preserve CMR since this characteristic is directly related to the mismatch in the associated time constants. Close component matching is also required to preserve the integrator response to e_2, because a canceling pole-zero pair exists in the response to this signal. Performance is also limited by the input offset voltage and offset current of the amplifier, as expressed by

$$\epsilon = \frac{1}{RC} \int V_{os} \, dt + \frac{1}{C} \int I_{os} \, dt + V_{os}$$

Noninverting integration of a single signal is also possible within the differential integrator if the e_1 input is grounded.

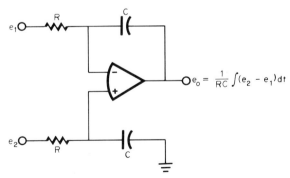

Fig. 7.5 Integration and subtraction are combined by the differential integrator.

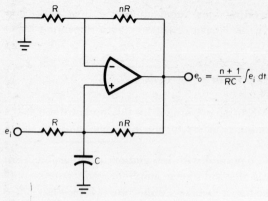

Fig. 7.6 Bootstrapping a simple RC integrator results in a noninverting integrator with accurate response.

Noninverting integration can also be performed with the bootstrapped RC integrator of Fig. 7.6.[4] The integral approximation of the input RC elements is linearized by means of bootstrapping feedback from an operational amplifier. Essentially, this circuit consists of the current source of Fig. 2.15 with a capacitive load. For the ideal integrator response, the current in C must equal e_i/R. However, the current supplied by e_i through R equals the desired level only when capacitor voltage is zero. As the capacitor voltage builds from zero, an amplified equivalent of this voltage is developed at the circuit output. The latter voltage feeds back a response correction current through the nR bootstrap resistor to keep the capacitor current at the desired e_i/R level. Then, the capacitor signal voltage and its amplified version at the output are related to e_i by the integrator response indicated.

While this circuit provides noninverting integration without the need for the precise capacitor match required with Fig. 7.5, it does still depend upon critical resistor matching, and dc errors have greater effects. Output offset error, or integrator time drift, is

$$\epsilon = (n+1)\left(\frac{1}{RC}\int V_{os}\, dt + \frac{1}{C}\int I_{os}\, dt + V_{os}\right) + nRI_B$$

This integrator drift is further amplified if careful resistor matching is not provided since the circuit employs positive feedback. As with any bootstrapped circuit, the positive feedback must be precisely controlled to maintain the bootstrap correction and yet avoid self-sustaining signals. For an integrator, self-sustaining signals are most likely dc signals because dc signals receive highest amplification. As a result, excess bootstrap feedback can cause the integrator of Fig. 7.6 to rapidly drift to saturation. To avoid this, resistor matching must be accurate, and source resistance must be considered in this matching. Source resistance adds to the input resistor, increasing the positive feedback.

7.2.2 Extending integrator time constants

A wide range of integrator gains or time constants can be achieved through the appropriate choice of the integrator resistor and capacitor. However, the practical ranges of resistances and capacitances usable in integrators is limited by parasitic characteristics of these elements and by their mounting environment. Very large time constants are restricted by the characteristics of high-value resistors and capacitors. Large capacitance values are available only with less ideal characteristics and with large physical size. Resistors of extremely high value are less precise because of parasitic leakage resistances, parasitic capacitances, and fabrication limitations. To maintain accuracy with large time constants, the integrator resistor is replaced with a tee network composed of lower value resistors as in Fig. 7.7. By making R_2 small compared with R_1, the equivalent resistance of the tee network R_{eq} is made much larger than those of the resistors used. Parasitic capacitances and leakage resistances have much less effect on these smaller resistors, and they are more readily available as precision components. Parasitic capacitance and resistance across the tee network remain significant, however; so careful component mounting is still required. Lower input resistance results with the tee network, but otherwise circuit performance is like that of the basic integrator.

A lower limit is imposed on the time constant of the basic integrator by parasitic capacitances and by the minimum acceptable circuit input resistance. To maintain reasonable integrator input resistance, the resistor used cannot be made too low in value. Further, extremely low capacitance values are not available because of error from parasitic capacitances. To make possible lower time constants, the integrator capacitor is replaced with a tee network as in Fig. 7.8. By making C_2 large compared with C_1, the equivalent capacitance of the tee is made

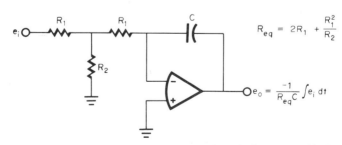

Fig. 7.7 For high integrator time constants without the limitations of high-value resistors and capacitors, the integrator input resistor is replaced with a tee network.

much smaller than that of either capacitor. Parasitic capacitances shunting C_1 and C_2 have much less effect than on a capacitor of value equal to C_{eq}. Integrator performance is otherwise the same as though the tee network were a capacitor of equivalent value. Parasitic capacitance across the tee remains significant; so careful component mounting is still required. Inte-

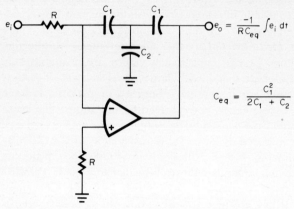

Fig. 7.8 For low integrator time constants without the limitations of low-value resistors and capacitors, the integrator capacitor is replaced with a tee network.

grator time drift is accentuated with low-value capacitors and with low time constants; so amplifier input offset voltage and current are more critical.

7.3 Differentiators

While not so accurate as integration, differentiation with operational amplifiers is also simple. In addition to its basic differentiation function, the operational amplifier differentiator can be simultaneously used for addition, subtraction, and noninverting response, as will be described. Also described are techniques appropriate for very large or very small differentiator time constants.

7.3.1 Summing and noninverting differentiators Addition and subtraction are readily combined with differentiation by using summing and feedback networks like those of summing and differencing amplifiers. Noninverting operation results with some of these and with a bootstrapped RC differentiator. To combine addition with differentiation, signals are summed at the input of the basic differentiator as in Fig. 7.9. Weighted gains are established for the various input signals through the use of different input capacitor values. Each capacitor is accompanied by a gain-limiting resistor needed to retain frequency stability with the differentiator connection,[1] and these resistors are inversely weighted.

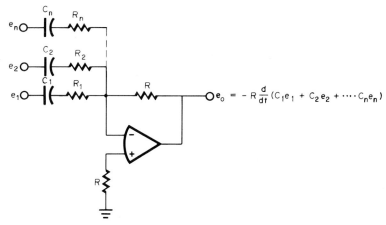

Fig. 7.9 Addition as well as differentiation is performed by a differentiator with input summing elements.

While any number of input signals could be summed in this manner with an ideal differentiator, increasing the number of summing elements lowers feedback factor. As a result, gain accuracy and noise performance are degraded in the summing differentiator. Amplifier input voltage noise is amplified by the combined gains of the various signal connections. For a differentiator, the added amplification of noise can be a serious limitation because the noise of the basic differentiator alone is significant. This is a result of the differentiator gain magnitude response, which increases linearly with frequency such that high-frequency noise is greatly amplified. The resulting output noise can mask low-frequency signals, which receive far less amplification.

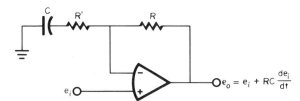

Fig. 7.10 Signal connection to the noninverting input of a differentiator results in noninverting operation and addition of the signal to its derivative.

Where it is desired to add a signal to its derivative, a simple differentiator variation is available. In this case, the input signal is connected to the noninverting amplifier input, rather than to the input capacitor, as in Fig. 7.10. This connection also provides noninverting differentiation and the high input impedance of noninverting operational amplifier connections. Performance is otherwise that of the basic differentiator. Further signal summation can be performed by the circuit through the connection of signal summing elements to the inverting amplifier input as in Fig. 7.9.

The differential differentiator of Fig. 7.11 is formed by adding a network that is identical to the feedback network to the noninverting input. In each network a gain-limiting resistor R' is added to the basic elements in order to ensure frequency stability.[1] Below the frequency of this gain limit, the response of the circuit is approximated by the expression in the figure. As expressed, the differential differentiator develops the difference between the time derivatives of two signals. In this way the output of a floating source can be differentiated without its common-mode signal.

Several sources of error limit the differential differentiator response. Common-mode signals introduce one of the errors of this circuit. The circuit common-mode rejection is limited by that of the operational amplifier and by mismatch in the attenuations of the two networks. Close matching of the two networks is also required to produce a pole-zero cancellation in the actual circuit response. Gain error also limits the accuracy of differentiator circuits such as this one. Ideally, a differentiator would display a continuously increasing gain with increasing frequency, which cannot of course be attained. Gain-bandwidth limitations of the operational amplifier greatly limit differentiator response. When the stabilizing gain limit resistor R' is used, the differentiator response approximation ends at $f = 1/2\pi R'C$.

Noninverting differentiation can be performed with the differential differentiator by simply driving the e_2 input with the e_1 input grounded. Alternatively, a differentiator can be built without the phase inversion of the conventional differentiator circuit.[5] With the noninverting configuration the gain magnitude has a positive polarity that would otherwise require addition of an inverter amplifier to the conventional differentiator. As shown in Fig. 7.12, the noninverting differentiator consists of a rudimentary RC differentiator that is bootstrapped and buffered by an amplifier. By itself the input RC network produces only a rough approximation to a

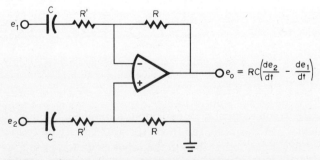

Fig. 7.11 Differentiation and subtraction are combined by the differential differentiator.

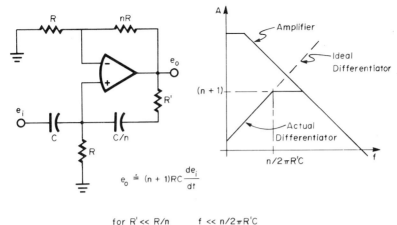

for R' << R/n   f << n/2πR'C

Fig. 7.12 Bootstrapping of a rudimentary RC differentiator results in a precision differentiator without the phase inversion of the conventional operational amplifier realization.

differentiator response. To correct the response, a positive feedback signal is supplied through the capacitor C/n. The amplifier amplifies the corrected differentiator signal to produce the output response

$$e_o \doteq (n+1) \, RC \, \frac{de_i}{dt}$$

for $R' \ll R/n$ and $f \ll n/2\pi R'C$.

The result is the differentiator response illustrated, which is much like that of the conventional differentiator.[1] To ensure frequency stability, the response must deviate from that of the ideal differentiator before its intercept with the open-loop response of the amplifier. Otherwise, the combined phase shifts of the operational amplifier and the differentiator feedback would be sufficient to induce oscillation. Stability is ensured by choice of R' and control of the net positive feedback. As indicated, R' is chosen to limit the gain-bandwidth product of the differentiator $n(n+1)/2\pi R'C$ to less than about one-third that of the amplifier. The positive feedback will cause oscillation, if it becomes greater than the negative feedback because of deviations in the component ratio matches indicated. Ratio mismatches between R and nR or between C and C/n will both cause gain error and/or oscillation that can be removed by trimming R or nR. Note that C and C/n act as capacitive loading to the amplifier, and this can also lead to oscillation.

7.3.2 Extending differentiator time constants

A wide range of differentiator gains or time constants can be achieved through the appropriate choice of the differentiator resistor and capacitor. However, a practical range of resistors and capacitors usable in differentiators is limited by the parasitic characteristics of these elements and their mounting environment. Very large time constants are restricted by the characteristics of higher value resistors and capacitors. Large capacitance values are available only with less ideal characteristics and with large physical size. Resistors of extremely high value are less precise because of parasitic leakage resistances,

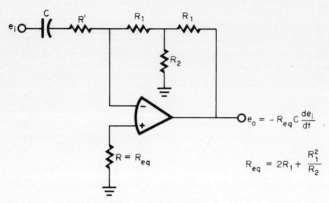

Fig. 7.13 For high differentiator time constants without the limitations of high-value resistors and capacitors, the differentiator feedback resistor is replaced with a tee network.

parasitic capacitances, and fabrication limitations. To maintain accuracy with large time constants, the differentiator resistor is replaced with a tee network composed of lower value resistors as in Fig. 7.13. By making R_2 small compared with R_1, the equivalent resistance of the tee network R_{eq} is made much larger than those of the resistors used. Parasitic capacitances and leakage resistances have much less effect on the smaller resistors, and they are more readily available as precision components. Parasitic capacitance across the tee network remains significant, however; so careful component mounting is still required.

A lower limit is imposed on the time constant of the basic differentiator by parasitic capacitances and the minimum acceptable circuit feedback resistance. Lower values of feedback resistance require greater amplifier output current, and this current is limited. Further, extremely low capacitance values are not available because of error from parasitic capacitances. To make possible lower time constants, the differentiator capacitor is replaced with a tee network as in Fig. 7.14. By making C_2 large compared with C_1, the equivalent capacitance of the tee network is made much smaller than that of either capacitor. Parasitic capacitances on C_1 and C_2 have much less effect than on a capacitor of value equal to C_{eq}. Differentiator performance is otherwise the same as though the tee

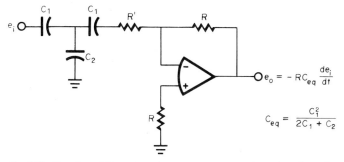

Fig. 7.14 Very low differentiator time constants are achieved without the limitations of low-value capacitors and feedback resistors when the differentiator capacitor is replaced with a tee network.

network were a capacitor of value C_{eq}. Parasitic capacitance across the tee remains significant; so careful component mounting is still required.

7.4 Multipliers and Dividers

Numerous techniques have been derived for analog multiplication and division.[1,2] With the availability of low-cost operational amplifiers, the logarithmic technique has become one of the best choices because it permits the use of the highly accurate logarithmic current-voltage characteristic of bipolar transistors. As will be discussed, this logarithmic approach can be extended to provide exponential powers and roots of signals along with multiplication and division.

Logarithmic conversion is readily performed with operational amplifiers to achieve multiplication and division as represented in Fig. 7.15. Input signals are converted to logarithmic equivalents, which are then added or subtracted as appropriate. Antilogarithmic conversion of the resulting combined signal then produces an output signal equal to the product

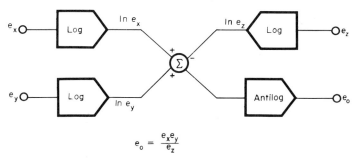

Fig. 7.15 Electronic multiplication and division can be performed through logarithmic conversions followed by addition and subtraction and then antilogarithmic conversion.

188 Designing with Operational Amplifiers

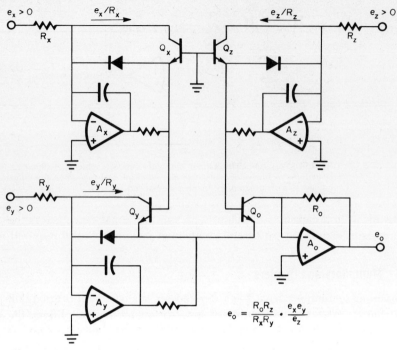

Fig. 7.16 The logarithmic multiplier employs logarithmic amplifiers, summation, and an antilogarithmic amplifier to perform the multiplication and division of Fig. 7.15.

or quotient of the various input signals. A practical realization of this technique is illustrated in Fig. 7.16. In this circuit, basic logarithmic amplifier structures are repeated in the A_x, Q_x and A_z, Q_z combinations. A similar connection is used for the A_y, Q_y logarithmic amplifier except that Q_y is connected to Q_x to permit addition of the associated logarithmic signals.

Similarly, antilogarithmic transistor Q_o is interconnected with Q_y and Q_z to complete the desired addition and subtraction of logarithmic signals. The result of the transistor interconnections is an emitter-base voltage established on the antilogarithmic transistor equal to

$$v_{beo} = v_{bex} + v_{bey} - v_{bez}$$

Each of the individual emitter-base voltages is related to the logarithm of its respective transistor collector current by[1]

$$v_{be} = \frac{KT}{q} \ln \frac{i_c}{\alpha I_s}$$

If the transistors are matched and thermally coupled, each will have the same forward transfer ratio α, the same reverse saturation current

I_s, and the same temperature T. Then combination of the previous two equations defines the collector current in the antilogarithmic transistor in terms of the input signal currents by

$$i_{co} = \frac{i_{cx} i_{cy}}{i_{cz}}$$

Relating these signal currents to the associated signal voltages yields the transfer function of the multiplier/divider as

$$e_o = \frac{R_o R_z}{R_x R_y} \frac{e_x e_y}{e_z}$$

$$= \frac{e_x e_y}{e_z}, \quad \text{for } R_o = R_x = R_y = R_z$$

The result is the classical multiplier/divider transfer relationship with numerator terms controlled by two input signals and a denominator term controlled by a third. With this response, simultaneous multiplication and division are possible through signal connections to the three inputs. Where only multiplication is desired, a fixed reference voltage is connected in place of e_z. Similarly, for divider operation e_x or e_y is replaced by a reference voltage. The circuit sensitivities to each of the three signal inputs can be controlled through the choice of its associated input resistor.

Performance of the multiplier/divider of Fig. 7.16 is characterized by high accuracy but is limited by single polarity restrictions and signal-dependent bandwidth. Of primary concern in multiplier/divider applications is response nonlinearity, which is minimized in this logarithmic approach. Nonlinearity due to transistor mismatch does not result for this circuit, as it commonly does for other multiplier circuits that make use of the exponential emmiter-base current-voltage characteristic. In this case, transistor mismatch creates only gain error, and nonlinearity is reduced to that associated with transistor emitter ohmic resistance. The ohmic contact and bulk resistances of the transistor emitters develop emitter-base voltage components that vary linearly with emitter current, instead of logarithmically as desired. As a result, multiplier response exhibits a nonlinearity because of the emitter ohmic resistance, but this can be reduced to 0.1 percent of full scale.

Gain error in the circuit response results from several sources, but it is immune to the thermal drift that is inherent in logarithmic amplifiers. Overall circuit gain error results from transistor mismatch and resistor mismatch. Gain drift is associated with corresponding drift mismatches. While the high thermal drift of the logarithmic conversion might be expected to contribute high drift for the multiplier/divider, it does not. A matching thermal drift exists in the antilogarithmic conversion, which compensates the multiplier/divider thermal drift to within the degree of match. To ensure this compensation, the four transistors must be thermally sinked to avoid temperature differentials.

Further circuit error results from the input dc errors of the operational amplifiers. The input offset voltages and bias currents of A_x and A_y result in signal feedthrough when only one of the signals e_x or e_y is zero. Ideally, one of these signals at zero would ensure zero output independent of the other signal, but the dc input errors offset the input zero, permitting signal feedthrough. For divider applications, the input dc errors of A_z will cancel the effect of small e_z signals, giving the effect of zero in the denominator of the divider response. Output dc errors, then, become very large as e_z approaches the offset level. However large, the associated divider-mode offset errors are far less than encountered when using a multiplier in an operational amplifier feedback loop for division.[1] From the remaining amplifier A_o straightforward output dc errors result from input offset voltage and bias current.

Because the logarithmic-antilogarithmic transistors of Fig. 7.16 can conduct for only one polarity of collector current, the input signals must be restricted to positive voltages. Negative voltages would reverse-bias the transistors, and clamp diodes are added to prevent damage from associated emitter-base breakdown. Unipolar input signals limit operation to one polarity quadrant of the four possible with two input variables. To permit both polarities of input signals, continuing transistor biases can often be established. Such biases do alter the circuit transfer function; so additional signal corrections are required. Four-quadrant multiplier operation is achieved by summing e_z into A_x and A_y with correction signals of e_z, e_x, and e_y summed into the A_o input.

With the logarithmic multiplier/divider approach, circuit bandwidth is signal-dependent, just as it is for logarithmic amplifiers.[2] Signal current variations in a logarithmic amplifier transistor modulate the dynamic emitter resistance of that transistor. The result is a signal-dependent

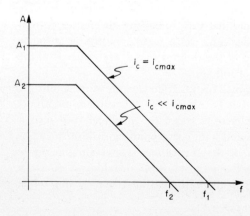

Fig. 7.17 Logarithmic amplifier bandwidth varies with the impressed signal such that the logarithmic multiplier bandwidth varies with its three input signals.

transistor gain in the logarithmic amplifier feedback loop, which combines with the gain of the operational amplifier. As illustrated in Fig. 7.17 the signal-dependent logarithmic amplifier gain results in a signal-dependent bandwidth. To accommodate the added gain of the logarithmic transistors, added phase compensation is required. This is provided in Fig. 7.16 by the feedback capacitors and amplifier output resistors.[2] This phase compensation must be selected for the maximum feedback gain occurring when the logarithmic transistor currents are maximum.

The logarithmic multiplier/divider technique can be extended to combine power and root functions with multiplication and division. In this way, multiple functions can be performed with one circuit, which is commonly called a *multifunction* circuit. When using logarithms, power and roots of a number are found by multiplying or dividing the logarithm of the number by another number and then finding the antilogarithm. To perform this operation electronically, a logarithmic signal is amplified or attenuated by a constant factor prior to antilogarithmic conversion. Such an operation is performed by the multifunction circuit of Fig. 7.18. As with the previous logarithmic multiplier, the emitter-base voltages of logarithmic transistors are added or subtracted to develop the emitter-base bias for the antilogarithmic transistor Q_o. By means of this addition and subtraction of logarithmic signals, the multiplier/divider function is again achieved.

However, for this multifunction circuit the logarithmic signals to be combined are attenuated or amplified because of the R_1 and R_2 voltage dividers. By virtue of the $R_1, n_1 R_1$ divider, the signal at the output of A_z is made $n_1 + 1$ times the difference between v_{bey} and v_{bez}. Since these emitter-base voltages represent logarithms of signals, increasing them by a factor raises the associated antilogarithm to a power equal to that factor. Similarly, the $R_2, n_2 R_2$ voltage divider attenuates the signal presented to the antilogarithmic transistor by Q_y and Q_z. This attenuation of logarithmic signals results in an antilogarithmic output effect equivalent to the taking of the root of that output.

Analysis reveals the nature of the power and root function achieved. Summing signals defines the emitter-base voltage established on the antilogarithmic transistor as

$$v_{beo} = v_{bex} + \frac{n_1 + 1}{n_2 + 1}(v_{bey} - v_{bez})$$

Each of these emitter-base voltages is logarithmically related to a transistor collector current; and if the transistors are matched, the above equation becomes

$$i_{co} = i_{cx}\left(\frac{i_{cy}}{i_{cz}}\right)^m \qquad m = \frac{n_1 + 1}{n_2 + 1}$$

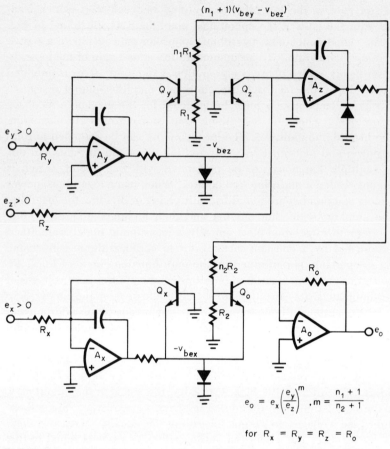

Fig. 7.18 Power and root functions, as well as multiplication and division, are performed by this multifunction extension of the logarithmic multiplier.

Relating these currents to their associated signal voltages results in the multifunction transfer equation

$$e_o = \frac{R_o}{R_x} e_x \left(\frac{R_z}{R_y} \frac{e_y}{e_z}\right)^m$$

$$= e_x \left(\frac{e_y}{e_z}\right)^m \quad \text{for } R_o = R_x = R_y = R_z$$

With this transfer function, positive and negative powers and roots of signals are possible in addition to multiplication and division. Power functions are achieved by choosing resistors for $n_2 = 0$ and $n_1 > 0$ through use of the $R_1, n_1 R_1$ divider. For root taking, n_1 is made zero and n_2 is made greater than zero using the $R_2, n_2 R_2$ voltage divider. Negative powers or roots result from interchanging connections to the e_y and e_z inputs.

Response accuracy with the multifunction circuit is controlled by the same factors described for the logarithmic multiplier and by the effects of the voltage dividers. Nonlinearity is introduced by transistor ohmic emitter resistance but not by transistor mismatch. Such mismatch introduces only gain and offset errors, as with the logarithmic multiplier, but the attenuators of the multifunction circuit increase the significance of these errors. For this reason, the range of practical exponents is limited to approximately 5 to $1/5$. Response bandwidth varies with signal amplitude, as described with the previous circuit, and the same phase-compensating elements are used here. Also used with the multifunction circuit are clamping diodes that limit transistor reverse emitter-base voltage to prevent damage from reverse breakdown.

7.5 Trigonometric Functions

Electronic computation of trigonometric functions has historically been performed with complex circuits that produce piecewise linear approximations to the functions.[4] Such circuits approximate the desired input-to-output function with a series of linear segments chosen for the best-fit correspondence to a function. A great number of such segments are required to make the actual response accurately fit the desired function. However, with the multifunction circuit described in the preceding section, trigonometric-function approximation for moderate accuracy requirements is simplified.[6] Using the multifunction circuit with noninteger exponents, power-series expansions for trigonometric functions can be approximated within 1 percent by a very abbreviated series.

Described below are multifunction implementations of sine, cosine, and arctangent functions.

For the sine function, a power-series expansion is

$$\sin x = x - \frac{x^3}{3!} + \frac{x^5}{5!} - \frac{x^7}{7!} + \cdots$$

This expansion could be realized electronically by using a series of multiplier/dividers to develop the terms with their integral exponents. However, fewer terms are required for a given approximation accuracy if nonintegral exponents available with a multifunction circuit are used.

With one multifunction circuit and a difference amplifier, two terms are realized for about 0.2 percent full-scale error in sine approximation using the connection of Fig. 7.19. Actual circuit response is expressed by

$$e_o = \frac{R_2}{R_1} \left(e_i - \frac{e_i^{2.83}}{6.28(E_R)^{1.83}} \right)$$

$$\doteq \frac{R_2}{R_1} E_R \sin \frac{e_i}{E_R} \quad 0 < e_i < \frac{\pi}{2} E_R$$

where e_i/E_R represents a radian angle. Angular response scaling is achieved through the choice of a voltage reference E_R, and amplitude scaling by choice of a difference amplifier gain R_2/R_1. This approximation to the sine-function power series remains accurate only for one angular quadrant as expressed. Additional accuracy limitations are, of

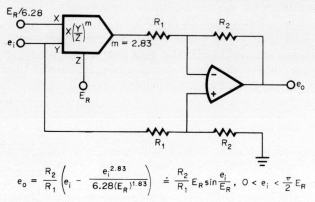

Fig. 7.19 Sine-function approximation is simplified with a multifunction circuit that provides a nonintegral exponent for an abbreviated power-series expansion.

course, imposed by the precision with which the multifunction circuit and difference amplifier develop the approximating expression.

A similar multifunction approximation yields a 1 percent cosine approximation. Again, a multifunction circuit and a difference amplifier are combined as illustrated in Fig. 7.20. In this case, the multifunction

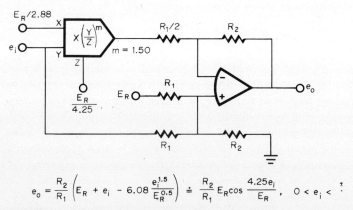

Fig. 7.20 A 1 percent cosine approximation is developed with the multifunction circuit of Fig. 7.18 and a difference amplifier.

exponent is made 1.50, and a reference voltage is summed into the difference amplifier. The result is a transfer function of

$$e_o = \frac{R_2}{R_1}\left(E_R + e_i - 6.08\,\frac{e_i^{1.5}}{E_R^{0.5}}\right)$$

$$\doteq \frac{R_2}{R_1}\,E_R \cos\frac{4.25 e_i}{E_R} \qquad 0 < e_i < \frac{\pi E_R}{8.5}$$

As before, scaling is achieved through the choice of E_R for angular gain and R_2/R_1 for amplitude. Approximation accuracy is limited to one angular quadrant for this truncated series.

One other useful trigonometric-function approximation performed with

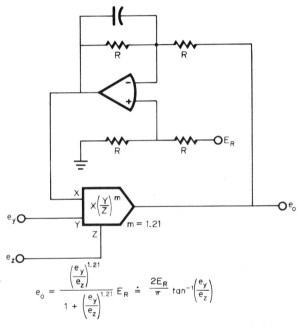

Fig. 7.21 An arctangent response is achieved with a multifunction circuit and a feedback difference amplifier.

a multifunction circuit is the arctangent. This function is frequently applied in rectangular-to-polar coordinate conversion. For the arctangent approximation a difference amplifier is connected in a feedback configuration with a multifunction circuit as in Fig. 7.21. The result is a circuit response of

$$e_o = \frac{\left(\frac{e_y}{e_z}\right)^{1.21}}{1 + \left(\frac{e_y}{e_z}\right)^{1.21}} E_R \doteq \frac{2E_R}{\pi} \tan^{-1} \frac{e_y}{e_z}$$

While the error of this approximation is less than 1 percent of full scale, the absolute error can be quite large for low input signal levels because of the input offsets of the multifunction circuit and the amplifier.

Because of the feedback loop of the arctangent converter, the phase shifts of the amplifier and multifunction circuit combine, increasing the possibility of circuit oscillation. This potential is furthered by the signal-induced bandwidth variations of the multifunction described in the previous section. Low signal levels result in significantly reduced bandwidth as illustrated before in Fig. 7.17, and associated with the lower bandwidth is increased phase shift at lower frequencies. For the arctangent converter of Fig. 7.21 the feedback signal presented to the x input becomes small when e_y is large compared with e_z. To ensure frequency stability under this condition, the feedback bypass capacitor shown is added.

7.6 Specialized Functions

With combinations of the preceding computing circuits an almost limitless variety of mathematical computations can be performed electronically. Two of the more common requirements are vector magnitude computation and rms conversion. Both of these functions can be realized through straightforward implementations of their describing equations; however, simpler specialized realizations are available.

For vector magnitude computation, the desired circuit response is

$$e_o = \sqrt{e_1^2 + e_2^2}$$

Applications of such a computation include rectangular-to-polar coordinate conversion and force-vector resolution. To implement this response, two multipliers could be used to square e_1 and e_2 followed by a summing amplifier and a square-root circuit. A far simpler circuit realization of this function results from use of feedback in an *implicit solution*.[4] With the implicit solution technique it is first assumed that the desired function has been realized; then the function is fed back for use in the actual derivation.

By this means, the circuit of Fig. 7.22 derives the vector magnitude function using a multiplier/divider and two operational amplifiers. Use of feedback from the output in this way replaces a multiplier and a square-root circuit of the straightforward implementation with an operational

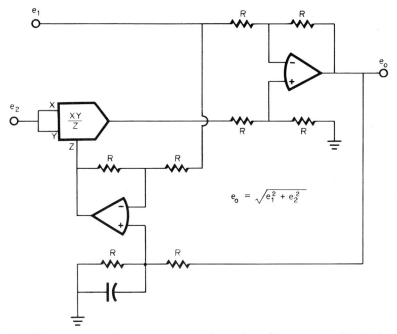

Fig. 7.22 Vector magnitude computation is made simpler and more accurate through the use of internal circuit feedback.

amplifier. In addition, accuracy and input voltage range are improved by avoiding direct squaring of the input signals. If e_1 and e_2 were actually squared, the resulting signals would cover a far greater dynamic range and, therefore, be much more sensitive to offset errors. With the approach of Fig. 7.22 no square-term signals actually occur, except where divided by other terms, and internal signal dynamic ranges are greatly reduced. This same feature permits application of much larger input voltages without developing internal circuit voltages that would saturate component outputs.

A disadvantage of using internal feedback for this vector magnitude computation is the combined phase shift of elements in the feedback loop. To preserve frequency stability in the presence of this feedback phase shift, phase compensation in the form of the bypass capacitor shown may be required. If so, the resulting circuit response can be significantly slower than achieved with the straightforward implementation mentioned above.

An implicit solution can also be used to simply rms conversion providing the response

$$E_o = \sqrt{\overline{e_i^2}}$$

198   Designing with Operational Amplifiers

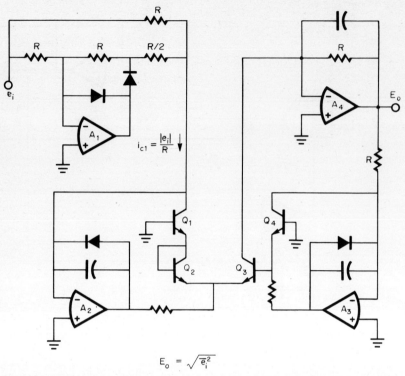

Fig. 7.23 Logarithmic signal processing and internal circuit feedback result in rms-to-dc conversion without need for squaring and square-root circuits.

Direct derivation of this response would require a multiplier for squaring and a filter for averaging, followed by a square-root circuit.[2] Instead, a logarithmic technique with internal circuit feedback can provide the rms-to-dc conversion with four operational amplifiers as in Fig. 7.23. In this circuit, squaring results from doubling a logarithm, averaging is provided by a filter capacitor, and the square-root response is realized through feedback.

Required with this logarithmic approach is an absolute-value conversion of the input signal, since the logarithmic converting transistors can accept only one polarity of collector current. That conversion is made by A_1 with its feedback elements, which are diode-switched to maintain a unipolar current supply to the logarithmic transistors of A_2. If Q_1 and Q_2 were replaced by a resistor, A_1 and A_2 would essentially form the common absolute-value circuit of Fig. 5.17.

In unipolar form the signal current can be processed by the logarithmic elements to produce rms conversion. That current is conducted by two logarithmic transistors Q_1 and Q_2 to develop the doubled logarithmic voltage needed for the squaring operation. The resulting logarithmic voltages are impressed on antilogarithmic transistor Q_3, along with a logarithmic voltage developed on Q_4 by feedback from the circuit output. Voltage drive to the antilogarithmic transistor is then

$$v_{be3} = v_{be1} + v_{be2} - v_{be4}$$

For matched transistors these emitter-base voltages are logarithmically related to currents such that

$$i_{c3} = \frac{i_{c1} i_{c2}}{i_{c4}}$$

where

$$i_{c2} = i_{c1} = \frac{|e_i|}{R}$$

Relating these collector currents to signal voltages defines the circuit response by the desired rms function:

$$E_o = \sqrt{\overline{e_i^2}}$$

This logarithmic approach to rms conversion has requirements and constraints similar to those of the logarithmic multiplier/divider for Fig. 7.16. Clamping diodes are added to the feedback paths of A_2 and A_3 to prevent the damaging emitter-base breakdown of the transistors. Because the transistors add gain in these feedback paths, additional phase compensation is required in the form of amplifier feedback capacitors and series output resistors. This added gain varies with signal current level, producing bandwidth variation for the rms converter like that described for the referenced multiplier/divider.

REFERENCES

1. G. Tobey, J. Graeme, and L. Huelsman, *Operational Amplifiers: Design and Applications*, McGraw-Hill Book Company, New York, 1971.
2. J. Graeme, *Applications of Operational Amplifiers: Third-Generation Techniques*, McGraw-Hill Book Company, New York, 1973.
3. D. Sheingold, Calculating Resistances for Sum and Difference Networks, *Electronics*, June 12, 1975.
4. G. Korn and T. Korn, *Electronic Analog and Hybrid Computers*, 2d ed., McGraw-Hill Book Company, New York, 1972.
5. J. Graeme, Bootstrapped RC Differentiator Performs Accurately without Phase Inversion, *Electron. Des.*, March 1, 1974.
6. D. Sheingold, Approximate Analog Functions with a Low-Cost Multiplier Divider, *EDN*, February 5, 1973.

8
DATA TRANSMISSION CIRCUITS

Much industrial electronic instrumentation, such as used in process control, is characterized by remote signal monitoring. As the distance increases between a signal monitor and its signal processing system, a special set of instrumentation problems is encountered. Simply the quantity of wire needed can be a major system expense, especially if the wire must be shielded against the noise pickup to which long lines are vulnerable. Longer lines can also introduce resistances that can absorb a significant portion of signal voltages, unless heavier gauge wire is used, and great distances between signal monitors and the signal processor result in differences in "ground" potentials approaching several hundred volts. Each of these instrumentation difficulties can be overcome with one or more of the two-wire transmitters or voltage-to-frequency converters of this chapter.

8.1 Two-Wire Transmitters[1]

For hardwired remote monitoring the minimum set of wires required to transmit a signal consists of the signal line and its ground return. A third wire would be normally required to supply power to the monitor, unless maintenance limited battery power is used. Two-wire transmitters keep the number of required wires to the minimum set of two by combining the signal and power lines. Only the power-supply line and its return are required. The signal is transmitted in the form of the supply current drain.

As modeled in Fig. 8.1, a two-wire transmitter is a voltage-to-current converter that transmits its signal current on its own supply lines. The output current consists of a quiescent level I_0 and a signal current related to the input signal e_i by a transconductance g_m. In process control systems I_0 and g_m are generally set for current ranges of 4 to 20 mA or 10 to 50 mA. An output voltage is derived from this current by merely connecting a load resistor R_L in series with the return line.

Note that the input signal must be referenced to one of the two lines and not to its own ground reference. This requires that e_i be supplied from a floating sensor. If the sensor has a separate ground return, a current will circulate between the two ground connections and create an error voltage on R_L. Sensors such as thermocouples are readily floated. Other sensors that require bias can be connected in bridge circuits that are biased from the two-wire transmitter for the desired floating connection. This is illustrated later. Where the input signal must have its own ground return, an isolating two-wire transmitter (to be described) can be used.

In addition to reducing the number of wires required for signal monitoring, two-wire transmitters permit the use of less expensive wire. If the signal is transmitted as a voltage, shielding is often needed to limit noise pickup, and larger diameter wire is necessary to reduce error from signal voltage drops on the line resistance. However, signals in current form are nearly immune to noise voltage pickup; so shielding is not needed. Further, line resistance does not reduce a signal current as it does a signal voltage; so smaller gauge wire is adequate. Generally, a twisted pair can be used instead of a more costly shielded cable.

Two-wire transmitter circuit realizations vary with the monitor requirements. Described here are a basic circuit, an improved-accuracy configuration, a bridge input form, a modulated-carrier type, and an isolating two-wire transmitter.

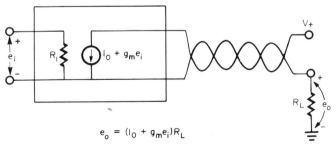

$$e_o = (I_0 + g_m e_i) R_L$$

Fig. 8.1 To transmit a signal, a two-wire transmitter produces a signal-dependent supply current drain in response to a floating signal voltage.

8.1.1 General-purpose circuits

The basic circuit consists of an amplifier biased from a floating zener power supply as in Fig. 8.2. Forming the amplifier are a noninverting connected operational amplifier and current-boosting transistors Q_1 and Q_2. A positive input signal will be amplified by a gain of $1 + R_2/R_1$ to develop a current in the amplifier load resistor R_3. This current and the amplifier feedback current are returned through D_4 to the transmitter load R_L. Relating this combined signal current to the input signal e_i is a transconductance of

$$g_m = \frac{R_1 + R_2 + R_3}{R_1 R_3}$$

Low-frequency accuracy of this transconductance is limited by resistor tolerance, amplifier loop gain, and by signal-dependent quiescent currents. Both the quiescent current of the amplifier I_{Q-} and of the floating supply I_S vary with the output signal e_o. That signal directly varies the drain-source voltage of Q_3, modulating I_S slightly. Similarly, the associated amplifier output signal is accompanied by current modulation

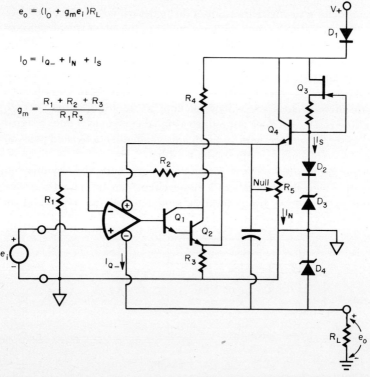

Fig. 8.2 Two-wire transmitter operation is achieved with an amplifier that supplies load current in response to an input signal and with a derived floating power supply.

within the amplifier. In addition, current drawn from the amplifier output directly lowers I_{Q-}. To limit that effect a Darlington current booster is used. Combined transconductance error from these signal-dependent quiescent currents is typically 0.3 percent of full scale. At higher frequencies the amplifier bandwidth limitation will similarly limit g_m.

Bias for the amplifier is derived from a floating zener diode power supply to limit feedback effects from the output swing e_o. Since the amplifier is bias-referenced to the output terminal, the output swing is present at the amplifier negative supply terminal. This swing is absorbed by current source Q_3 and emitter follower Q_4 rather than by the amplifier, because the floating positive supply is also referenced to the output terminal. Impressed across the amplifier power-supply terminals is a voltage essentially equal to that of the two zener diodes, as long as the voltage drop of D_2 counteracts that lost to the emitter-base junction of Q_4. This derived supply voltage can also be used to provide the required floating bias for a resistance bridge as illustrated with a later circuit. The zener voltage is chosen large enough for amplifier bias and swing requirements, but not so large as to limit the transmitter output e_o. Limiting of e_o will occur when it would exceed that portion of the external supply voltage V_+ that is not required by the floating supply. Protection from damage by reverse connection or transients of the external supply is provided by D_1.

To set the quiescent output current I_O at its desired level, a null current I_N from R_5 is added to the quiescent currents I_{Q-} of the amplifier and I_S of the floating supply. This gives

$$I_O = I_N + I_{Q-} + I_S$$

If I_O is to be only 4 mA, a low-quiescent-drain operational amplifier and low-current zener diodes are desirable. A low-quiescent-drain amplifier also helps reduce the thermal variations in I_O.

All three components of I_O vary with temperature and represent a major source of circuit error. The quiescent current of an operational amplifier will typically vary 20 percent over the amplifier operating temperature range. To minimize the quiescent current variation of the floating supply, its bias FET can be set at its nominal zero temperature coefficient point, but this is approximate. The null current varies with the zener voltage drift, which is generally increased by the need to maintain low zener current. Added to these quiescent current drifts is the effect of the input offset voltage drift of the operational amplifier. That voltage drift is multiplied by the circuit g_m to create further I_O drift. Combined thermal variations result in an output current drift of about $(0.03 + 0.02\ g_m)\ \%$ F.S./°C (where F.S. stands for *full scale*).

A number of circuit refinements can be made to reduce the thermal-

and signal-induced changes in I_0. Rather than stabilizing each of the components of I_0, they can be conducted through the amplifier load resistor. The total current in this resistor is precisely controlled by the amplifier feedback; so any change in quiescent current will be compensated for by feedback adjustment of that load resistor current. Both thermal- and signal-induced I_0 variations are thereby compensated for, reducing drift and transconductance errors.

One such feedback-controlled circuit realization is that of Fig. 8.3. Again the two-wire transmitter consists of an amplifier, formed with A_1, and a floating power supply, now developed with A_2. Circuit quiescent currents are conducted through the amplifier load resistor R_3. This connection, however, requires feedback and bias modifications. Since the negative side of the floating supply is now connected to the emitter of the current booster Q_1, that emitter does not swing with respect to the floating circuitry. Instead, the signal swing appears at the opposite end

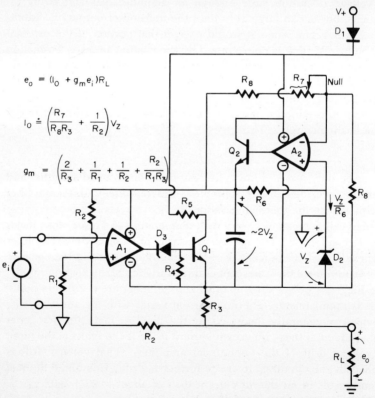

Fig. 8.3 Two-wire transmitter accuracy is improved by returning all quiescent circuit currents as part of the feedback controlled current in the amplifier load R_3.

of R_3, and feedback is returned from that point. This effectively inverts the phase of the feedback signal requiring feedback to the noninverting input of A_1.

Also because of the connection of the negative supply point to Q_1, dc level shifting must be performed at the output and input of A_1. Output level shifting is provided by zener diode D_3, which ensures an amplifier output above its saturation voltage level. Input saturation due to dc current in the R_2 feedback resistor is prevented by the R_2 input bias resistor connected to the positive supply.

Otherwise, the amplifier operation is similar to that previously described. A positive input signal e_i is amplified by a gain of $2 + R_2/R_1$, resulting in a controlled signal voltage on the amplifier load R_3. Current from R_3 and the R_2 feedback resistor is returned through the transmitter load resistor R_L to develop the output voltage e_o. Describing this input-to-output transfer is a transconductance

$$g_m = \frac{1}{R_1} + \frac{1}{R_2} + \frac{2}{R_3} + \frac{R_2}{R_1 R_3}$$

The accuracy of this transconductance at low frequencies is essentially determined by resistor tolerances. Signal-induced variations in quiescent currents that would create additional error are now conducted through R_3, where total signal is feedback-controlled. Some signal-related variation exists in the current supplied by the R_2 bias resistor. This results from variation in the floating supply with the power-supply rejection error of A_2 and with the signal current from R_1 flowing in the reference D_2. Fortunately, the supply rejection error is quite small for frequencies commonly of interest. Also, the signal-induced change in the reference diode voltage alters both positive and negative supply voltages to produce largely counteracting changes on the two R_2 resistors. As long as the R_2 resistors are large, signal-induced transconductance error is negligible.

To realize this low transconductance error, it was necessary to reduce the output resistance of the floating supply from that provided by the previous circuit. In Fig. 8.3 this is achieved by connecting the output pass transistor Q_2 in the feedback loop of an operational amplifier A_2. With the amplification provided by A_2, only one zener diode D_2 is needed. Note that D_2 is biased by R_6 from the output of this simple voltage regulator, rather than from the external supply V_+. Thus, the regulator stabilizes the bias for its own reference diode, avoiding the need for a current source to absorb signal swing.

Output voltage from this floating power supply is adjusted by the null potentiometer to set the transmitter quiescent output current I_0. The potentiometer causes the gain of A_2 to be slightly greater than 2, making the magnitude of the positive supply greater than that of the negative.

Because of this difference the current supplied through the R_2 input bias resistor develops a voltage on the R_2 feedback resistor that exceeds the negative supply voltage. This forces a voltage on R_3 and sets a quiescent output of

$$I_O \doteq V_Z \left(\frac{R_7}{R_8 R_3} + \frac{1}{R_2} \right)$$

As mentioned, the thermal drift of I_O is greatly reduced by including the quiescent currents of the circuit in the feedback-controlled path. Remaining I_O drift is associated with that of the zener reference, the input offset voltage V_{OS1} of A_1, and that amplifier input bias current I_{B1}. Zener drifts can be significant unless the diode is chosen for a low temperature coefficient at the lower current available for its bias. Thermal drift of V_{OS1} is multiplied by the circuit transconductance and produces the dominant drift at higher gain levels. Generally, the drift of I_{B1} is negligible compared with the feedback current. Combined, these drift sources result in

$$\frac{\Delta I_O}{\Delta T} \doteq \left(\frac{R_7}{R_8 R_3} + \frac{1}{R_2} \right) \frac{\Delta V_Z}{\Delta T} + g_m \left(\frac{\Delta V_{OS1}}{\Delta T} + \frac{R_2}{2} \frac{\Delta I_{B1}}{\Delta T} \right)$$

With care this drift can be controlled to $(0.003 + 0.02 g_m)$ % F.S./°C.

Each of the preceding two-wire transmitters is illustrated with an unbiased, floating signal source, such as is encountered with thermocouples. Transducers that require bias, such as variable resistance types, must have a floating bias that is referenced to the transmitter circuitry. Only then will the floating transmitter circuitry monitor the transducer signal separated from the transmitter output signal and from ground potential differences. Fortunately, the required floating bias is available from the floating power supply derived to bias the transmitter circuitry.

Either of the previous circuits of Fig. 8.2 or 8.3 can be used with a biased transducer, and the choice is determined by accuracy requirements. In addition to having superior response accuracy, the circuit of Fig. 8.3 has a more precise floating supply available for transducer bias. To use this circuit with a resistance bridge, it is modified as in Fig. 8.4. The bridge bias is set by the R_9 and $R_9/2$ resistors for an approximate transducer current of V_Z/R_9. To bias the inputs of A_1 within their common-mode voltage range, the $R_9/2$ resistor is added in series with the bridge. An unbalance in bridge currents is produced by the null potentiometer for an output null current of

$$I_O \doteq \frac{V_Z R_2 R_7}{2 R_9^2 R_3}$$

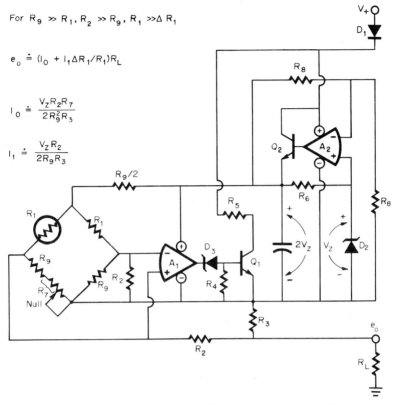

Fig. 8.4 Two-wire transmitter monitoring of biased transducers is facilitated by the available floating power supply of the transmitter.

While the above expression is approximate, the actual current is quite stable. As with the preceding circuit, I_0 is under the feedback control of A_1. Variations in I_0 are essentially those caused by the thermal variations in the zener voltage and the resistances expressed above. This thermal variation can be readily limited to 200 ppm/°C.

Changes in transducer resistance further unbalance the bridge currents to produce an output signal. Again, the difference current flows in the R_2 resistors, producing a differential signal that is matched on R_3 through feedback action. The currents associated with these signal voltages flow through the transmitter load R_L to produce an output voltage of

$$e_o \doteq \left(I_0 + I_1 \frac{\Delta R_1}{R_1}\right) R_L$$

where

$$I_1 \doteq \frac{V_Z R_2}{2 R_9 R_3} \quad \text{for } R_9 \gg R_1,\ R_1 \gg \Delta R_1,\ R_2 \gg R_9$$

Response conformance to this expression is primarily limited by the approximation of the expression. Because of bridge response nonlinearity, the transmitter response deviates from the above approximation where the approximation conditions listed becomes less valid. However, the inherent circuit accuracy does permit operation to within 0.1 percent of the exact response relationship. Since many variable resistance transducers lack this degree of accuracy in their own response, the approximate expression can often be used with confidence that accuracy is not significantly degraded by the two-wire transmitter.

8.1.2 Specialized two-wire transmitters

More specialized two-wire transmitters can be used to further reduce wiring or to accommodate a nonfloating signal source. These are a modulated-carrier type that permits common use of one wire pair by many monitors and an isolated circuit configuration that accommodates grounded signal sources. Reduced wiring with a modulated-carrier two-wire transmitter is made possible by the current mode output. Theoretically, any number of signal currents can be summed on a common line. If these currents are modulated carriers of differing frequencies, they can be reseparated at the line end. As a result, the wiring required to monitor numerous remote signals can be reduced to a single twisted pair.

Modulated-carrier output is fairly easily achieved with a bridge-sensing two-wire transmitter. As expressed with the previous circuit, the output of such a configuration is proportional to the product of the bridge unbalance and the bridge bias voltage. By replacing the bridge bias with a carrier signal, a transmitter output is developed that is the carrier modulated by the bridge unbalance. One such circuit is that of Fig. 8.5. Basically, this circuit resembles that of Fig. 8.4 with the addition of the Wien-bridge oscillator of Fig. 6.2 for driving the transducer bridge. The result of the associated carrier signal E_c is a modulated-carrier output of

$$e_o \doteq \left(I_0 + I_1 \frac{\Delta R_1}{R_1}\right) R_L$$

where

$$I_1 \doteq \frac{E_c R_2}{(R_8 + R_9/2) R_3} \quad \text{for } R_9 \gg R_1,\ R_2 \gg R_9,\ R_1 \gg \Delta R_1$$

Added to this ac output signal is a dc bias of approximately 30 mA as set by the unbalance between R_2 and R_6. These resistors also create an ac unbalance that is counteracted by the null potentiometer in setting I_0. In this case I_0 represents the output carrier at null, which can be set for a 4 mA amplitude. The 30 mA dc bias permits increasing the output carrier amplitude to 20 mA for a consistent output range.

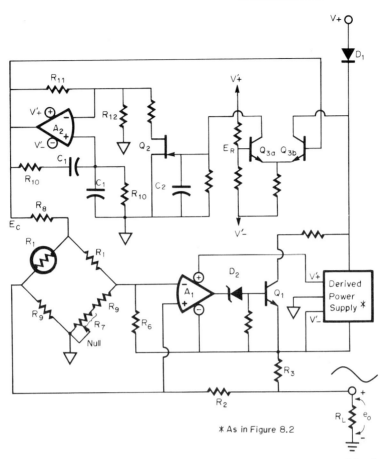

Fig. 8.5 Wien-bridge oscillator drive of a transducer bridge results in a modulated-carrier two-wire transmitter for multiple use of a single wire pair.

Response conformance to the above output expression is limited by the approximations of the expression, as described for the last circuit. Circuit-oriented error is primarily determined by the amplitude stability of the Wien-bridge oscillator. For this reason an AGC loop with gain is used to control the oscillator amplitude. The basic oscillator consists of A_2, the R_{10}, C_1 Wien bridge, and the gain-setting feedback of R_{11} and R_{12}. Frequency is set by the Wien bridge at $f = 1/(2\pi R_{10} C_1)$. Amplitude is determined by critical control of the oscillator gain, and AGC is applied to finely tune that gain through FET Q_2. The FET acts as a voltage-controlled resistor that shunts R_{12} to raise gain. If the gain set by R_{11} and R_{12} is slightly below that required for oscillation, then the gain added by the FET shunting will control the amplitude of oscillation.

The actual amplitude is detected by the comparator formed with Q_{3a} and Q_{3b}. When the amplitude is below the reference level E_R at the base of Q_{3a}, that transistor is never turned on. No signal is then supplied to the FET gate, and gate-source bias is zero. At this bias the FET resistance is minimum for maximum oscillator gain. That gain level causes amplitude to increase until it reaches the reference level. Then the negative sine-wave peaks turn Q_{3a} on so that it applies bias to the FET gate. Under this bias the FET begins to turn off, lowering gain until an equilibrium amplitude is reached. The result is a 0.2 percent amplitude stability for similar two-wire transmitter accuracy.

Each of the preceding two-wire transmitters accepts only floating signal sources. While the floating requirement is compatible with most transducers, it is not suitable for more general instrumentation. For transmission of signals between widely separated instrumentation systems, the two-wire transmitter is again attractive for the reduction in wire expense it permits. However, the systems to be interconnected by the transmitter will have separate ground references which can be separated by large ground potential differences. Such potential differences can readily exceed the voltage handling capability of the two-wire transmitter, as well as overshadow the signal to be transmitted. The signal would be lost in ground circulation error currents that result with separate input and output ground connections, as described with Fig. 8.1.

To accommodate the separate signal and load ground references, without the difficulties described above, an isolating two-wire transmitter can be used. By means of an isolated coupling, the ground difference signal is rejected and ground circulating error currents are blocked. Only the desired signal is transmitted by the isolated coupling. One implementation of an isolating two-wire transmitter is shown in Fig. 8.6. It consists of a signal amplifier that drives an isolating LED (light-emitting diode)-phototransistor coupler and a dc-to-dc converter that powers the amplifier from the transmission wires. Alternatively, the amplifier can sometimes be powered from the system supplying the input signal, to eliminate the dc-to-dc converter.

Operation of the isolating two-wire transmitter is similar to that of previous circuits except for the isolated coupling and linearizing feedback of the LED-phototransistor couplers.[2] A positive input signal causes the output of the operational amplifier to supply current to the LEDs. This induces currents in the output phototransistor Q_1 and in its feedback counterpart Q_2. To satisfy feedback requirements, the current in Q_2 must be related to the input signal by the current e_i/R_1. If the couplers are matched, the same current will flow in the output phototransistor Q_1.

Current from Q_1 is doubled by Q_{3a} and Q_{3b} to accommodate the less than unity transfer efficiencies of the couplers and the dc-to-dc converter.

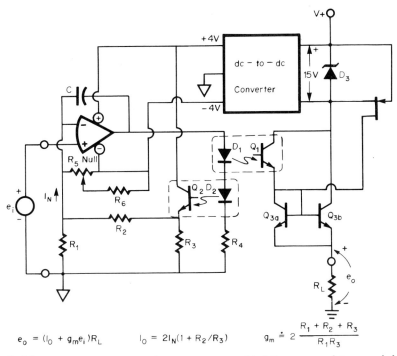

$$e_o = (I_0 + g_m e_i)R_L \qquad I_0 = 2I_N(1 + R_2/R_3) \qquad g_m \doteq 2\,\frac{R_1 + R_2 + R_3}{R_1 R_3}$$

Fig. 8.6 A two-wire transmitter with input-output ground isolation accommodates grounded signal sources.

From Q_1, current flows through Q_{3a}, which then biases Q_{3b} at an equal current. The result is a current gain of 2, and greater gain can be achieved by adding ratioed resistors in series with the emitters of Q_{3a} and Q_{3b}. This gain boosts that of the isolating couplers above the less than unity gain of most LED-phototransistor couplers. As a result, the power efficiency required of the dc-to-dc converter is lowered. Greater current is still required from the dc-to-dc converter output than is available to its input, but this is permitted by a large decrease in voltage from converter input to output. Transmitter output current not required by the dc-to-dc converter is conducted by D_3. This zener diode regulates the converter input voltage. To ensure circuit turn-on, a current is initially supplied to the output circuitry by the FET shown. Following turn-on, no current is supplied by the FET so long as its pinchoff voltage is less than the zener voltage of D_3.

As a net result of circuit operation, a current is supplied to the transmitter load for an output voltage of

$$e_o = (I_0 + g_m e_i)R_L$$

where

$$I_O = 2 I_N \left(1 + \frac{R_2}{R_3}\right)$$

$$g_m \doteq 2 \frac{R_1 + R_2 + R_3}{R_1 R_3} \quad \text{for } R_5 \gg R_1$$

Accuracy of this response is limited by about a 1 percent error from the mismatch, instability, and noise of the LED-phototransistor couplers. The couplers must be matched over the full signal dynamic range; so reduction of nonlinearity is difficult. Further, the time stability of the matching is degraded by the decay in transmission efficiency to which these couplers are prone. These couplers also have a high noise content to their output currents, resulting in another limit to signal sensitivity.

8.2 Voltage-to-Frequency Converters[3]

Voltage-to-frequency converters, VFCs, simplify analog-to-digital conversion for remote data conversion, digital voltmeters, and medical instrumentation.[4] These versatile converters produce an output frequency that is proportional to an input voltage with errors as low as 0.01 percent. Precise voltage control over frequency with simple circuitry is provided by the high feedback gain and linearity of operational amplifiers in a variety of circuits described below. These circuits provide output dynamic ranges of 1,000:1 and 10,000:1. With the narrower range circuits, full-scale errors can be limited to 0.4 and 0.1 percent. For the wider range circuits, full-scale errors from 0.03 to 0.01 percent can be achieved.

In the applications mentioned above the most convenient VFC output signal form is a pulse train because such a signal is readily decoded by a digital counter. Fig. 8.7 represents this approach in an analog-to-digital converter. As an analog-to-digital converter, this technique offers a truly monotonic, although slow, response. Slow conversion results from the

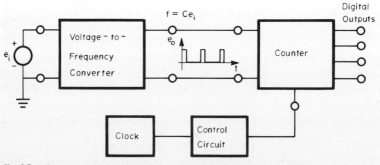

Fig. 8.7 The output of a voltage-to-frequency converter can be decoded by a counter in an analog-to-digital converter configuration.

need for a counting interval that is at least as long as the period of the lowest frequency signal. For VFC output frequencies extending down to 1 Hz, the counting interval must be at least 1 s. Such conversion times are, nevertheless, compatible with many remote data acquisition and instrument requirements. Faster conversion, as well as improved accuracy, can be attained with the input ranging generally employed with digital voltmeters and other instruments. Input ranging reduces the dynamic range required of the VFC, and operation can be restricted to the higher frequencies, where shorter counting intervals are adequate. To add input ranging to the VFCs presented below, it is only necessary to use switchable summing resistors on the input amplifier.

Input ranging is not readily adapted to remote data conversion applications, but VFCs have many other benefits for such applications. With a VFC mounted at a remote sensor or subsystem, the signal monitored is converted to a form suitable for more economic lines. The lines then required are simpler than needed for either analog or digital signal transmission. Line loss, distortion, and noise that directly affect analog signals have far less effect on the frequency-encoded information from a VFC; so less sophisticated wiring is suitable. Even noise large enough to introduce error in the decoding counter is averaged out if the counting interval is large compared with the period of the VFC output signal. Offsetting line voltage drops can be removed from the VFC signal with simple isolation techniques. The pulse train signal is readily transmitted across isolating transformers or optical couplers. This feature makes VFCs well suited to medical monitoring, where even small ground voltage differences represent a hazard.

Conventional analog-to-digital converters also provide the signal transmission benefits listed above, but the associated circuitry and often the wiring are more complex than for the VFC approach. The parallel bit output signals from an analog-to-digital converter must be converted to serial form to avoid separate lines for each bit. Even with the parallel-to-serial conversion, a separate line is needed to couple the system clock pulse train to the analog-to-digital converter, and that line is also subjected to the noise environment. With the VFC approach the clock signal is usually needed only with the decoding counter at the receiving end of the line.

The voltage-to-frequency converter appropriate for a given data conversion application is determined primarily by accuracy and conversion time requirements. Accuracy is governed by the errors and thermal drifts of the offset, gain, and linearity of a VFC. Described below are moderate-precision VFC circuits having total errors of approximately 0.4 and 0.1 percent and high-precision circuits with total errors of 0.03 to 0.01 percent. The greater the circuit accuracy, the wider is its useful dynamic range. As discussed before, the lower extreme of the dynamic range used determines conversion time. If only the upper portion of a VFC dynamic range is utilized, shorter conversion times are possible. Also, conversion times can be reduced by setting the circuits illustrated for higher frequency operation if greater error can be accepted.

8.2.1 Moderate-precision configurations For applications requiring error no greater than 0.4 percent of full scale, a VFC can be built with one operational amplifier as in Fig. 8.8. This circuit is basically a zener-controlled relaxation oscillator formed with an integrator and a level-detecting RESET switch. A constant negative voltage input to the integrator causes the output to increase linearly until the RESET switch is triggered. Triggering occurs when the output voltage is large enough to forward-bias the zener diode and the emitter-base junction of Q_1. Then, Q_1 drives Q_4 on, and positive feedback between these transistors rapidly increases their currents. As a result, a voltage pulse is generated at the emitter of Q_1 to turn on another positive feedback coupled pair Q_2 and Q_3. Current from the latter transistors overrides the signal current i_i and discharges the capacitor for reset. Discharge continues until the capacitor voltage drops below a level equal to the emitter-base voltage of Q_2 and the saturation voltage of Q_3. At that point, these transistors turn off, and the capacitor begins recharging to repeat the cycle.

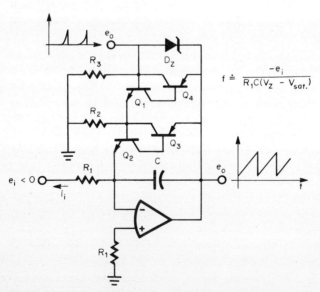

Fig. 8.8 A single-amplifier voltage-to-frequency converter is formed with an integrator and a voltage-sensing RESET switch.

During each cycle, the change in capacitor voltage ΔV is $V_Z - V_{sat}$. Cycle time Δt will be determined by the rate at which the input current i_i can charge C by ΔV, as long as the reset time is negligible. If the input signal remains constant during the cycle, C charges linearly, and $\Delta t = C\,\Delta V/i_i$. From this, the output, the signal frequency is found to be

$$f = \frac{1}{\Delta t} = \frac{-e_i}{R_1 C\,(V_Z - V_{sat})}$$

For this relationship to hold, the frequency of the input signal must always be much less than that of the output signal.

The actual response of the simple VFC of Fig. 8.8 deviates from that expressed above with offset, linearity, and gain errors. Output zero is offset by the input offset voltage and current of the amplifier as expressed by

$$\Delta f_O = \frac{V_{OS} + I_{OS} R_1}{R_1 C (V_Z - V_{sat})}$$

where Δf_O is the offset error. Note that the amplifier offset voltage V_{OS} can be adjusted to null all this error.

Response linearity is governed by the reset discharge time and amplifier gain error. While the discharge time t_D is negligible for low frequencies, it becomes increasingly significant as the frequency increases, and creates an error of $f^2 t_D$. Discharge time is generally limited by the rate at which the amplifier output can slew S_r; so $t_D = (V_Z - V_{sat})/S_r$. Significant at lower frequencies is the integrator gain error, and its effect on frequency is approximated by $f(V_Z - V_{sat})/A_O e_i$. Together the two nonlinearities produce a frequency error of

$$\Delta f_N = f(V_Z - V_{sat}) \left(\frac{f}{S_r} + \frac{1}{A_O e_i} \right)$$

where Δf_N is the nonlinearity error.

Gain error for the circuit of Fig. 8.8 is created by the tolerance variations of R_1, C, D_Z, Q_1, and Q_2. By adjusting R_1, the initial gain error can be made small, but a high gain-error drift restricts the improvement possible. High gain drift results from the thermal sensitivity of the RESET switch thresholds. Principal causes of this sensitivity are the thermal variations of the emitter-base voltages of Q_1 and Q_2. Matching these transistors helps to make their effects counteract, but their highly different currents and pulse operation limit improvement. Considering the various error sources, total circuit error under limited temperature excursion is approximately 0.4 percent of full scale for a 1 to 1,000 Hz frequency range.

To achieve improved VFC accuracy, the thresholds of the RESET

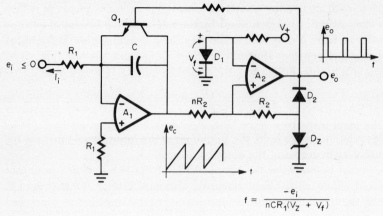

Fig. 8.9 Voltage-to-frequency conversion with 0.1 percent accuracy is provided by an integrator and a comparator reset circuit.

switch can be more precisely defined by a comparator as in Fig. 8.9. Once again, an integrator is used to convert the input voltage to a ramp that increases with time in proportion to the input voltage magnitude. The comparator uses a zener diode to control the high threshold and has hysteresis to also provide a low switching threshold.

Circuit operation is basically similar to that of the previous VFC. Negative input voltages cause the integrator output to increase until the high comparator threshold is reached at $e_c = nV_z + (n + 1)V_f$. At this point, the comparator output swings positive, turning on the RESET switch Q_1 to discharge C. Discharging continues until e_c reaches the low comparator threshold. That threshold merely equals V_f, because no feedback is supplied through the R_2 resistor to alter this switching point. Positive feedback is disconnected by D_2 during this RESET mode because the comparator output is positive. Because of D_1, the low threshold is held above zero. This is necessary since Q_1 cannot discharge C below the transistor saturation voltage.

From the above two comparator switching points, the change in capacitor voltage during each cycle is $n(V_z + V_f)$. The time required for this change to be developed by input current i_i defines the operating frequency, as long as the reset time is negligible. The charging time is $\Delta t = C \Delta e_c / i_i$ if i_i remains constant during the charging cycle. Frequency will then be

$$f = \frac{1}{\Delta t} = \frac{-e_i}{nCR_1 (V_z + V_f)}$$

Error performance of the VFC of Fig. 8.9 is characterized in terms of offset, linearity, and gain errors. Offset is produced by A_1 through its input offset voltage and current V_{OS1} and I_{OS1} and will be

$$\Delta f_0 = \frac{V_{OS1} + I_{OS1}R_1}{nCR_1 (V_Z + V_f)}$$

where Δf_0 is the offset error. By appropriate adjustment of V_{OS1}, this error can be largely removed.

Nonlinearity is introduced by the finite discharge time t_D and by the limited gain of A_1. These error sources become significant at opposite ends of the frequency range, where they produce increasing deviations from the ideal response. Expressions for these errors are similar to those described for the previous VFC and together result in a linearity error of

$$\Delta f_N = f \left[f t_D + \frac{n(V_Z + V_f)}{A_0 e_i} \right]$$

where Δf_N is the nonlinearity error. Reset time t_D will be determined by the slewing rate of the operational amplifiers or by the maximum current of discharge switch Q_1, whichever is more limiting.

Gain error is created by the tolerance variations of the R_1, C, R_2, nR_2, D_1, and D_Z components. Compensating adjustment of R_1 reduces this error to a level limited only by the thermal gain drift. Gain drift is greatly reduced from that of the last VFC by comparator setting of switching thresholds. Threshold drift is now essentially that of V_Z and V_f. By choosing the zener diode for a V_Z drift that counteracts that of V_f, the gain drift can be repeatedly reduced to 0.01%/°C. This permits far more accurate operation. Total error for operation under limited temperature excursion will be less than 0.1 percent of full scale over a 1 to 1,000 Hz range.

8.2.2 High-precision voltage-to-frequency converters

Higher precision VFC operation generally requires a correspondingly wider dynamic range for increased resolution. For 0.1 percent accuracy a resolution of 1 part in 1,000 is needed, and this requires a 1,000:1 dynamic range. Less dynamic range is suitable only if input range switching is performed, and such switching is not generally convenient in remote monitoring applications.

The previous VFC circuits provide a 1,000:1 range, but extension of this range is limited by conversion time and reset time. Conversion time places a limit on the minimum operating frequency, and for practical conversion times not exceeding 1 s the minimum usable VFC frequency is 1 Hz. Greater dynamic range must then come from extending the high-frequency end of the operating range. However, the high-frequency operation of the previous circuits is limited by the time required for resetting the integrator capacitor. As described, the reset time introduces a linearity error. To reduce this linearity error to the 0.01 percent consistent with a 10,000:1 operating range, the reset time must be no more than 0.01 percent of the minimum signal period. Typically, this would be 10 ns, and such short reset intervals are difficult to achieve and result in similarly short output pulses that would be absorbed by the capacitance of long lines.

Thus, extended dynamic range for higher precision requires capacitor reset by some means that does not introduce timing error. This is accomplished by supplying a controlled amount of resetting charge to the integrating capacitor, rather than the fixed reset voltage applied in the previous circuits. A controlled resetting charge can be applied at any time during a cycle without interrupting the integration of the input signal. With no separate reset time interval, the timing error is avoided.

For VFC operation the frequency with which the reset charge is supplied is made proportional to the input signal. This proportionality results if the reset charge is supplied each time the input signal charges the integrating capacitor to a reference voltage; the larger the signal, the more frequently the capacitor will recharge to that level. Such operation is illustrated in Fig. 8.10, where the integrator output rises to a reference level E_R at a rate determined by the current i_i from an input signal. At the reference level, a discharge current I_D is added to supply resetting charge for a time t_D. As this process repeats for a given i_i, the charging and discharging voltages are equal. Thus, the charging and discharging charges are equal and opposite:

$$Q_C = i_i t_C = -Q_D = I_D t_D$$

Then the oscillation frequency will be

$$f = \frac{1}{t_C} = \frac{i_i}{I_D t_D} = \frac{-e_i}{R I_D t_D}$$

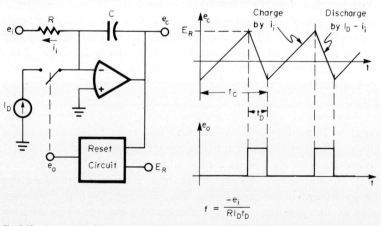

Fig. 8.10 Controlled charge reset makes the reset a normal part of a voltage-to-frequency converter cycle to remove the linearity error caused by a separate reset interval.

Data Transmission Circuits   219

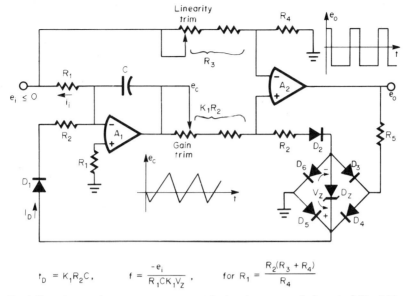

$$t_D = K_1 R_2 C, \qquad f = \frac{-e_i}{R_1 C K_1 V_Z}, \qquad \text{for } R_1 = \frac{R_2(R_3 + R_4)}{R_4}$$

Fig. 8.11 Voltage-to-frequency conversion with the charge-controlled reset of Fig. 8.10 is performed with an integrator and a signal-dependent comparator for a 10,000:1 dynamic range.

If I_D and t_D are constants above, the frequency is proportional to the input signal.

Two circuits are described below that make use of this controlled charge approach to extend dynamic range and reduce error to 0.03 and 0.01 percent. The first of these circuits is formed with an integrator and a resetting comparator as shown in Fig. 8.11. While these two elements of the circuit are similar to those of Fig. 8.9, the comparator is now arranged to provide a controlled discharge current I_D and a fixed discharge time t_D. This ensures a constant reset charge of $Q_D = -I_D t_D$, as required for the operation described above. Controlled discharge current is achieved by driving the integrator summing resistor R_2 with a fixed voltage from a zener diode. To produce a fixed discharge time, rather than a fixed discharge voltage, the input signal is coupled to the comparator, as will be described.

Application of a negative input signal induces the VFC oscillation. The negative signal drives the comparator A_2 to its positive output state, where it conducts a current through D_4, D_Z, and D_6. This establishes a voltage of $V_Z + V_{f6} - V_{f1}$ on the R_2 input resistor for a discharge current of

$$I_D \doteq \frac{V_Z}{R_2} \qquad \text{for } V_{f6} = V_{f1}$$

Because this current is made greater in magnitude than i_i, the integrator output is now a negative-going ramp. That ramp continues to the first comparator trip point. Only e_i controls this trip point because D_2 disconnects the comparator positive feedback when e_o is positive. As a result, the first trip point is at $e_c = e_i/K_2$, where $K_2 = (R_3 + R_4)/R_4$.

At this trip point, the circuit switches to the charging mode. The comparator output switches to its negative state, reverse-biasing D_1 to disconnect the discharging current. Current i_i from the input signal then charges the capacitor until the second trip point is reached. That trip point is controlled by both e_i and the positive feedback of the comparator. In this negative output state, D_2 is forward-biased to connect the positive feedback, and the diode bridge inverts the voltage presented by the zener diode. As a result, the same zener diode is used to establish the second trip point and the discharge current. This ensures the matching required for good linearity. In this state the zener diode results in a voltage $-V_Z -V_{f5} + V_{f2}$ at the R_2 feedback resistor of the comparator. This voltage and the input signal define the second comparator trip point at

$$e_c = K_1 \left(V_Z + \frac{e_i}{K_2}\right) + \frac{e_i}{K_2} \quad \text{for } V_{f5} = V_{f2}$$

From the two trip-point levels the operating frequency is defined. The difference in trip points fixes the change in e_c at

$$\Delta e_c = K_1 \left(V_Z + \frac{e_i}{K_2}\right)$$

Note that Δe_c, which is the reset voltage, is a function of e_i rather than a constant. It is this fact that makes the discharge time t_D a constant, as required for the linear response $f = e_i/RI_D t_D$. The discharge time is

$$t_D = \frac{\Delta e_c C}{I_D - i_i} = \frac{\Delta e_c C}{V_Z/R_2 + e_i/R_1}$$

If Δe_c were a constant, t_D would vary with e_i. However, Δe_c from above has a compensating variation with e_i, and

$$t_D = \frac{K_1 (V_Z + e_i/K_2) C}{V_Z/R_2 + e_i/R_1} = K_1 R_2 C \quad \text{for } R_1 = K_2 R_2$$

With t_D a constant the VFC response is linearly related to e_i by

$$f = \frac{-e_i}{R_1 C K_1 V_Z} \quad \text{for } R_1 = K_2 R_2$$

Performance of the VFC of Fig. 8.11 is determined by component selection and the resulting offset, gain, and linearity errors. While this can provide a 0.01 percent resolution, accuracy is limited to approximately

0.03 percent of full scale by the errors mentioned. Offset error is again created by the dc input errors of the integrator and will be

$$\Delta f_O = \frac{V_{OS1} + I_{OS1}R_1}{R_1 C K_1 V_Z}$$

where Δf_O is the offset error. Note that there is a value of V_{OS1} for which Δf_O is zero and the offset control of A_1 can be adjusted to reduce the frequency offset to approximately 0.1 Hz. Gain error results from the tolerances and drifts of R_1, C, D_Z, D_1, D_5, D_6, R_2, and the $K_1 R_2$ resistance. These errors are compensated for by adjustment of the gain trim potentiometer and can be reduced to about 0.003 percent.

Linearity is dependent upon the open-loop gain of A_1, the switching time of A_2, the $R_1 = K_2 R_2$ ratio, and the frequency sensitivity of two diode matches. Limited open-loop gain in A_1 results in an input error signal that introduces nonlinearity at low frequencies. This input error signal is greatest at low frequencies for an integrator such as A_1. It is also at low frequencies that VFC input signals are their smallest and, thereby, most sensitive to the amplifier input error signal. So, a high-gain operational amplifier is desired. At the high extreme of the VFC frequency range, the comparator switching time and the error in the $R_1 = K_2 R_2$ resistance ratio introduce nonlinearity. The comparator switching time imposes a frequency-sensitive change in hysteresis. Error in the K_2 resistance ratio degrades the compensation in t_D that corrects for presence of the input current i_i during the discharge period. Since i_i is greatest for high VFC output frequency, its effect on t_D is greatest there.

Increasing frequency also degrades two important diode voltage matches. Increased VFC frequency is associated with an increasingly frequent supply of reset charge. This makes for a greater duty cycle for matched diodes D_1 and D_6 and a lesser duty cycle for the matched pair D_2 and D_5. As a result, the diode temperatures vary, disturbing the match in forward voltage drops needed to maintain constant I_D and t_D. Best results are achieved if monolithic dual devices are used for the diode pairs. The combined nonlinearity can be reduced to approximately 0.02 percent of full scale by means of the linearity trim potentiometer. Added to this trimmed error level is a thermal drift of typically 0.006%/°C. Overall error for a 1 to 10,000 Hz range can be limited to 0.03 percent of full scale.

To achieve greater accuracy, the nonlinearity of the above circuit must be avoided. At the expense of an external clock supply, this can be achieved with a discharge time t_D that is precisely controlled by a clock signal. Again, the controlled charge reset described with Fig. 8.10 is applied for a linear response of $f = e_i/RI_D t_D$. However, the clock signal control permits 0.01 percent accuracy. The reset circuitry is controlled

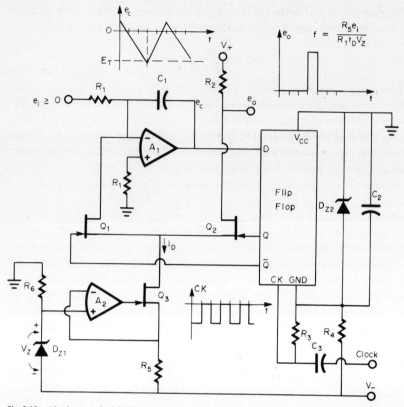

Fig. 8.12 Clock control of discharge time provides excellent linearity to voltage-to-frequency conversion.

by the clock as shown in Fig. 8.12, where the discharge current is supplied for one period of a precision clock signal. As a result, the frequency of the output signal will always be a subharmonic of the clock frequency, but the average of the output signal frequency will be that desired.

Once again, the integrator capacitor is charged by the input signal until the integrator output reaches a reference level. In this case, the reference level is the threshold voltage of a flip-flop. While this reference is not precise, it need only be constant during each cycle to ensure linear response. When the integrator output reaches the flip-flop threshold, it keys the flip-flop for switching at the next clock pulse. That clock pulse causes the Q output to switch to its low state while the $\overline{Q}$ output switches to high state. In this mode Q_1 is turned on to supply a discharging current, and Q_2 is turned off to provide a high circuit output state. By appropriate choice of I_D, the integrator output will be discharged beyond the flip-flop threshold before the next clock pulse arrives, and the circuit will then be switched back to its charging state. This operation fixes the discharge time at the precise period of the clock pulse. For a constant reset charge, a constant discharge current I_D is also needed. That current is supplied by a precision current source[2] formed with A_2 and Q_3, and $I_D = V_Z/R_5$. This gives the desired VFC response

$$f = \frac{R_5 e_i}{R_1 t_D V_Z}$$

As with the previous circuits, the performance limitations of the VFC of Fig. 8.12 are specified in terms of offset, gain, and linearity errors. Offset error is once more determined by the dc input errors of the integrator and will be

$$\Delta f_O = \frac{R_5(V_{OS1} + I_{OS1}R_1)}{R_1 t_D V_Z}$$

By means of the offset control of A_1, this error can be reduced to 0.001 percent of full scale. Gain errors are produced by the tolerances and drifts of R_1, C, D_{Z1}, R_5, and the clock pulse period. To adjust gain, the R_1 input resistor is trimmed. Nonlinearity is all but removed as a source of error for the resolution permitted by the 10,000:1 dynamic range. Variation in the discharge time now only produces gain error. The only residual error that varies with frequency to introduce nonlinearity is controlled to permit an overall circuit accuracy of 0.01 percent of full scale.

REFERENCES

1. J. Graeme, Improve Analog Data Transmission with Two-Wire Transmitters, *Electron. Des.*, January 5, 1975.
2. J. Graeme, *Applications of Operational Amplifiers: Third-Generation Techniques*, McGraw-Hill Book Company, New York, 1973.
3. J. Graeme, Use V/F Converters for Analog Data Transmission, *Electron. Des.*, April 1, 1975.
4. S. Conners, Voltage-to-Frequency Converters: A/D's with Advantages, *EDN*, June 5, 1974.

9
TEST AND MEASUREMENT CIRCUITS

Operational amplifier feedback provides precise control and simplified adjustment of signals for a wide range of test and measurement requirements. These include active and passive component testing, and measurement of numerous signal characteristics. Described are several more specialized circuits for transistor and operational amplifier testing. Other test circuits included are ohmmeters and capacitance meters. Incorporated for signal measurement applications are operational amplifier realizations of voltmeters, ammeters, frequency meters, and phase detectors. Concluding the chapter are electronic thermometer circuits.

9.1 Active Component Test Circuits

Precision testing of active devices is achieved through accurate control of test conditions and by isolation of the device under test from the loading effects of test monitors. With operational amplifiers, independent control of the test bias and signal voltages and currents is made possible by the high-gain feedback and buffering of these amplifiers. Also provided by the impedance buffering of operational amplifiers is the desired isolation of monitor loading. These qualities are applied to transistor and amplifier testing for numerous performance characteristics,[1,2] and additional test applications are described below.

9.1.1 Transistor test circuits Operational amplifiers have been used to simplify testing of many transistor characteristics.[1] Simplification results from the biasing, buffering, and ground referencing provided by operational amplifiers in this testing. These benefits are applied here to testing of more specialized transistor characteristics, including MOSFET threshold voltages, photocoupler efficiency, and transistor package thermal resistance.

The threshold voltage of an enhancement-mode MOSFET represents that gate-source voltage required to initiate the current flow in the device. Ideally, this is the voltage at which the FET source current changes from zero to a nonzero level; so measurement of a low current level is involved in testing this characteristic. To avoid measurement loading of this low current, operational amplifier buffering is beneficial. In addition, the buffer amplifier can be interconnected with the MOSFET in a manner that makes the test current independent of the developed gate-source voltage.

This is achieved by connecting the MOSFET in feedback around the operational amplifier as in Fig. 9.1. For this test the source current I_S is set only by bias voltage V_+ or V_- and resistor R. That current is not affected by the MOSFET gate-source voltage, because feedback removes that voltage from R by maintaining zero voltage between amplifier inputs. Feedback forces the gate of the device under test to that level from ground that causes the MOSFET to conduct the test current I_S. As long as this current is small compared with the rated MOSFET current and large compared with leakage, the amplifier output voltage will nearly equal the threshold voltage of the device under test. At that output the threshold voltage can be measured without introducing loading error to the test current.

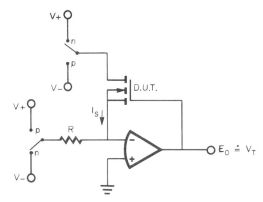

Fig. 9.1 MOSFET threshold voltage testing is simplified with the buffering and test current control provided by an operational amplifier.

Operational amplifier control of test signals can also ease testing of optical coupler transfer efficiency as in Fig. 9.2. A reference voltage E_R is accurately transformed into an LED drive current through the feedback of A_1. In maintaining zero voltage between its inputs, that amplifier conducts a diode current I_D that develops a voltage equaling E_R on R_1. The resulting phototransistor current is converted into a ground-referenced voltage by the current-to-voltage converter formed with A_2. This permits test result measurement with a ground-referenced voltmeter. From that measurement, the transmission efficiency is found by the relationship

$$\gamma = -\frac{R_1}{R_2}\frac{E_O}{E_R}$$

As expressed, the choice of R_1 and R_2 sets the test circuit gain, while E_R and R_1 set the test current. For sweep testing, E_R can be replaced by a ramp train, and the resulting output monitored on an oscilloscope.

Also simplified with an operational amplifier is thermal resistance testing of transistors. Thermal resistance is a major indicator of the power dissipation capability of a packaged transistor; so it is generally specified by the manufacturer. However, common specifications apply for either no heat sink or an infinite heat sink. Neither case really applies in practice, because some nonzero, finite heat sinking is always provided by the transistor mounting and heat sinking. Thus, thermal resistance testing of a transistor in its intended mounting environment is desirable for accurate prediction of transistor operating temperature. One means of performing this test is to merely measure the case temperatures of a test sample of mounted transistors and then to calculate junction temperature using published case-to-junction thermal resistance.

Alternatively for bipolar transistors, thermal resistance can be measured electrically in one step by using the transistor as its own temperature sensor. The emitter-base voltage of a silicon bipolar transistor varies

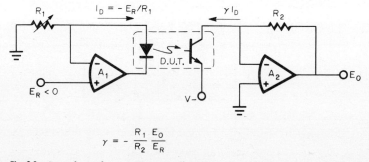

Fig. 9.2 Optical coupler transmission efficiency can be measured using operational amplifiers to develop a controlled LED current and to convert phototransistor current to a ground-referenced voltage.

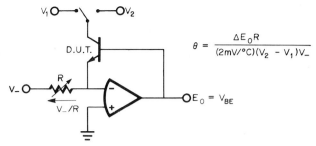

Fig. 9.3 To measure the thermal resistance of a bipolar transistor, collector voltage is switched, and the emitter-base voltage change is monitored to determine the resulting temperature change.

linearly with temperature by about -2 mV/°C. This voltage can be monitored to determine the temperature change induced by a given increase in power dissipation to then define actual thermal resistance. To implement this test, the circuit of Fig. 9.3 holds transistor emitter current constant while collector-emitter voltage is switched to change the transistor power dissipation. By switching the collector voltage from a low level V_1 to a higher level V_2, the transistor power dissipation is increased proportionally. The increase is determined by the collector-emitter voltage change and the emitter-current level set by R and V_-. Accompanying the change in power dissipation will be a variation in the transistor emitter-base voltage, which is established at the amplifier output. If the two output voltage levels are E_{O1} and E_{O2}, the thermal resistance is related by

$$\theta = \frac{\Delta T}{\Delta P} = \frac{(E_{O1} - E_{O2})R}{(2 \text{ mV/°C})(V_2 - V_1) V_-}$$

Some error is introduced in this measurement by the transistor base current loss and Early effect. Because some of the emitter current is conducted through the base, the power dissipation is slightly less than assumed above. Also, the variation in collector-base voltage directly alters the transistor emitter-base voltage through base-width modulation as described by the Early effect. Fortunately, both errors usually have a negligible effect on thermal resistance testing in comparison to the thermal resistance variations associated with the variables of transistor mounting.

9.1.2 Operational amplifier test circuits

A variety of test circuits have been developed for testing operational amplifiers.[1] Additional test circuits presented in this section provide measurement of thermal feedback error in monolithic operational amplifiers and permit simultaneous testing of five basic operational amplifier characteristics.[3] With the test circuit

228 Designing with Operational Amplifiers

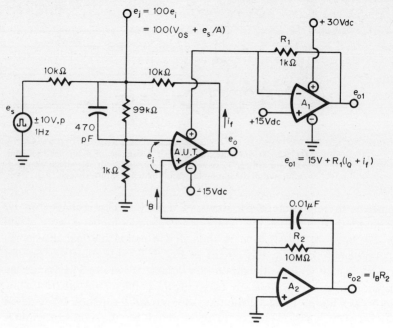

Fig. 9.4 Simultaneous testing of five basic operational amplifier characteristics is performed by this circuit.

shown in Fig. 9.4 signals are generated that provide measures of open-loop gain, offset voltage, input bias current, quiescent current, and output voltage swing. The various output signals can be separately monitored or processed and combined to produce a single pass-fail indication. For the open-loop gain indication the ac portion of the summing junction signal e_j is measured. This signal is an amplified replica of the operational amplifier input signal e_i, which in turn is related to the output signal e_o by the amplifier gain A. As a result, the monitored signal e_j is related to the gain as shown. With a peak-to-peak detector the ac portion of e_j can be converted to a gain-related dc voltage suitable for pass-fail examination by a voltage comparator. (A square-wave test signal is shown because it is simple to generate with a single amplifier; however, a sinusoidal test signal could be used.)

Input offset voltage V_{OS} is measured from the dc portion of e_j. Again, a comparator can be used to provide a pass-fail indication. If e_j is not filtered for this monitor, the amplitude of its ac component will introduce some error. But generally the high gain and moderate offset voltage levels of general-purpose operational amplifiers result in an ac component that is an order of magnitude smaller than the dc component; so filtering can often be eliminated to speed testing.

A measure of input bias current I_B is provided by the simple current-to-voltage converter formed with A_2. Flow of input bias current through R_2 creates a dc voltage at the output of A_2 for the comparison against a desired level. To avoid introduction of measurement error by A_2, its offset voltage should be nulled, and its input bias current should be much less than that to be measured. While this test only checks one of the two input bias currents, this is generally adequate where very low input offset current is not required. Any input transistor mismatch that would create excessive input offset current would very likely also create high offset voltage that would be detected by the V_{OS} monitor.

In a manner similar to the I_B measurement, quiescent current is monitored by A_1. As shown, A_1 develops a voltage from the positive supply current drain of the amplifier tested. This voltage is partly due to the feedback current i_f as well as I_Q. However, i_f is known for a given feedback resistor and output swing; so its effect can be corrected by offsetting the monitor limit.

Output voltage swing is checked to the level of the input signal, because the amplifier under test is basically in a unity-gain inverter configuration defined by the 10 kΩ resistors. If the output cannot swing the full range of the input signal, a large error signal will be introduced to e_j, and the gain monitor will detect this failure. By using comparator monitors with OR gating, a single pass-fail output indication can be produced for the five tests performed.

For monolithic integrated-circuit operational amplifiers, thermal feedback commonly introduces gain error that is best tested with a visual display. With its high thermal conductivity, low thermal mass, and small size, a monolithic chip readily conducts heat from dissipation at its output to its sensitive input. The thermal feedback introduces input offset voltage changes with the load current variations accompanying signal swings. These input voltage changes can override summing junction signals to alter the open-loop gain response in unexpected ways, making gain error prediction inaccurate. Careful amplifier layout limits the thermal gradients which cause thermal feedback signals, but some generally remain.

Illustrated in Fig. 9.5 are two input-output response curves that show ways in which thermal feedback can introduce gain error not anticipated from open-loop gain specifications. For open-loop conditions, the slope of the input-output response between output saturation levels represents the amplifier open-loop gain. This gain is typically tested and specified as an average for the rated output swing. As such, the response anticipated from the specified gain would be that of the dashed lines, and the actual response typically follows those lines in the absence of thermal feedback.

However, thermal feedback in monolithic operational amplifiers causes the input-output response to deviate from that anticipated at signal frequencies below a few hertz. At such low frequencies, the signal period is greater than the output-to-input thermal transfer time of the monolithic chip; so output circuitry heating will produce signal-related input offset voltage changes. At higher frequencies, the heating effect is averaged by the thermal time constant of the chip, and the associated input offset change does not follow the signal.

Where the thermal feedback effect does follow the signal, the induced input offset voltage change appears as a gain error, resulting in amplifier input-output transfer responses such as those shown by solid lines in Fig. 9.5. Both responses shown result in unexpected gain error. For the more common case of Fig. 9.5a the gain error introduced by thermal feedback is maximum when the output voltage is midway between zero and either saturation level. It is near these midpoints that power dissipation is maximized in amplifier output transistors when supplying current to a grounded load. At these points, the actual response has its maximum deviation from that anticipated. The significance of these deviations is that the associated closed-loop error can be several times that expected. Such gain error cannot be removed through feedback resistor adjustment because the error is signal-dependent rather than constant.

Another type of thermal feedback effect is represented by the solid response curve of Fig. 9.5b. In this case, the thermal feedback is positive and actually reverses the slope of the response curve. While this positive feedback will be overridden by negative feedback connected to the amplifier, the related closed-loop gain error will be opposite in sign from that anticipated. To remove this gain error, there must be an increase rather than a decrease in negative feedback.

To determine the error introduced by thermal feedback for a given type of monolithic operational amplifier, the input-output response can be

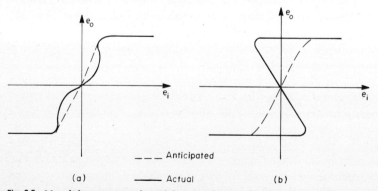

Fig. 9.5 Monolithic operational amplifier open-loop gain response can deviate from that anticipated because of internal thermal feedback.

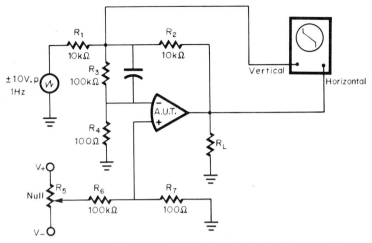

Fig. 9.6 Operational amplifier thermal feedback can be displayed on an oscilloscope using a low-frequency open-loop gain test circuit.

displayed on an oscilloscope as in Fig. 9.6. The basis of this test circuit is the common open-loop gain test configuration,[2] which connects the amplifier under test for a closed-loop gain of -1 through feedback resistors R_1 and R_2. To develop a larger, more measurable summing junction signal, a voltage divider consisting of R_3 and R_4 is added within the feedback loop. Using the resulting amplifier summing junction signal to drive the vertical input of the oscilloscope and the amplifier output signal to drive the horizontal input, an input-output response is displayed. That response will be rotated 90° with the input connected to the vertical axis, but this makes available the high gain of the vertical input to the low-level summing junction signal. Test signal frequency is chosen low enough to permit significant, signal-related feedback and lower than the frequency of the first amplifier open-loop response pole to avoid hysteresis in the display from the amplifier phase shift.

9.2 Ohmmeters

A basic ohmmeter consists of a voltage reference that develops a current in a resistor under test, and a meter that monitors that current. With operational amplifiers the accuracy and versatility of ohmmeters is extended in a variety of ways, as will be described. For conventional ohmmeter applications, operational amplifiers buffer the reference voltage and isolate meter resistance from the measurement circuit. In addition, operational amplifiers can be utilized to permit measurement of a given resistor embedded within a circuit.

9.2.1 Conventional ohmmeters

Operational amplifier realizations of ohmmeters can produce inverse, proportional, or null meter responses. Where a meter response inversely proportional to resistance and 2 percent error are acceptable, one operational amplifier serves to buffer the reference voltage and to isolate the resistor tested from meter resistance. These features are provided by the circuit of Fig. 9.7, which is derived from a controlled current-source configuration.[1] In this circuit a reference voltage E_R is established at the amplifier input by a zener diode with a switchable voltage divider incorporated for range switching. The reference E_R is buffered from the current drain of the test resistor R_x by the high input resistance of the operational amplifier. Amplifier feedback retains the test resistor voltage at a level equal to E_R. To do so, the amplifier drives an FET as required to supply the appropriate current to R_x. That current is also conducted through the FET to drive the meter. Meter resistance will not alter the current in R_x because of the isolation provided by the FET. The result is a meter current providing an accurate indication of the resistance of R_x by the relationship

$$R_x = \frac{E_R}{I_M}$$

Measurement accuracy for the ohmmeter of Fig. 9.7 is primarily limited by resistor, zener, and meter errors. Additional error is introduced by the amplifier input offset voltage and input bias currents. However, the

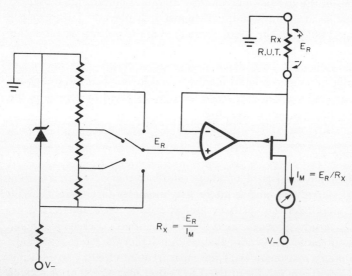

Fig. 9.7 Basic ohmmeter circuitry is improved with a controlled current source that buffers the reference and isolates meter resistance from the resistor under test.

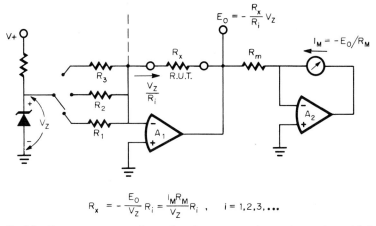

Fig. 9.8 Ohmmeter response that is directly proportional to resistance is provided by a two-amplifier configuration.

input offset voltage and its drift are small compared with the zener error and drift, and input bias current can be made negligible through the choice of amplifier. Error produced by resistor, zener diode, and meter inaccuracies can be removed by calibrating adjustments within the limits imposed by the zener voltage temperature coefficient and the meter nonlinearity. Such calibration is achieved through adjustment of the switched voltage divider. Iterative adjustment may be required since the adjustments interact. Residual error can typically be reduced to 2 percent of midscale.

Ohmmeter accuracy can be improved if response is made directly proportional to resistance, rather than inversely proportional as above. With the inverse proportionality the meter scale is greatly compressed on one end, making that portion of the meter movement unusable for accurate measurements. To make use of the full meter scale, an ohmmeter having meter drive directly proportional to resistance is implemented with two operational amplifiers as in Fig. 9.8. In this case, the unknown resistor R_x is connected so that it sets the amplification provided by an inverting amplifier A_1 to a reference voltage V_Z. The result is an output voltage E_O proportional to R_x. Also controlling the closed-loop gain of A_1 are its input resistors, which are switched for ranging.

A second amplifier A_2 is used to drive the meter if required. That amplifier converts E_O to a feedback current I_M, which is conducted through the meter. Alternatively, the voltage E_O can be monitored by a separate dc voltmeter. In either case, the ohmmeter response is directly proportional to resistance as expressed by

$$R_x = -\frac{E_O}{V_Z} R_i = \frac{I_M R_M}{V_Z} R_i \qquad i = 1, 2, 3, \ldots$$

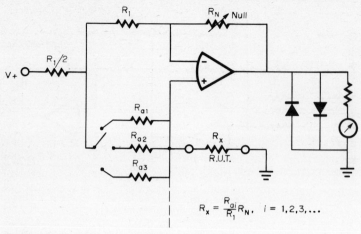

Fig. 9.9 Meter and reference voltage errors are removed from ohmmeter response with a null measurement.

Response error is dominated by meter nonlinearity, zener voltage deviations, and resistor tolerances as described for the previous circuit. Calibration adjustments of the A_1 input resistors reduce error to essentially the level of the meter nonlinearity. If the meter amplifier is not used and E_O is used as the circuit output, error can be further reduced to a level determined by the zener voltage temperature coefficient. This improved response accuracy does, however, remain vulnerable to the error of a monitoring voltmeter.

To avoid meter and reference voltage errors, null measurement techniques can be applied to ohmmeters as in Fig. 9.9. This circuit is a bridge-connected or difference amplifier for which zero output voltage results when the bridge is balanced. Balance is achieved through adjustment of the null potentiometer R_N while observing the meter for a zero indication. When the bridge is nulled, the resistance of R_x can be read from a turns counting dial on R_N considering the relation

$$R_x = \frac{R_{ai}}{R_1} R_N \qquad i = 1, 2, 3, \ldots$$

Null operation greatly improves ohmmeter accuracy. Meter error contribution is reduced to its offset, which is largely removed along with amplifier offset errors during calibration. For improved null resolution, full-scale meter indication is set for small amplifier output voltages, and this output voltage range is restricted by clamping diodes. Also removed is the error produced by reference voltage deviations, because the bridge bias introduces balancing voltages on the bridge in the null condition. One of the bridge elements is switched as shown for ranging. With the meter and reference errors removed, accuracy is primarily determined by the linearity of the null potentiometer movement. Other errors can be compensated for by calibrating adjustments to readily limit overall error to 0.1 percent of full scale.

9.2.2 Ohmmeters for embedded resistors Measurement of a resistor embedded in a circuit or in a network would usually require removal of that resistor. Otherwise other elements would shunt the resistor under test, producing measurement error. Removal of the resistor of interest is not only inconvenient but sometimes impossible if it is a hybrid resistor pattern or an element of a packaged network. To permit measurement of one element in a circuit, the feedback control of operational amplifiers can be used to prevent the shunting of other elements, such that the total test current flows in the resistor under test. In this way, measurement or trimming adjustments can be made for elements connected in closed loops or in tee networks that have no access to the junction of the tee.

For measurement of a resistor connected in a closed loop, the ohmmeter of Fig. 9.10 is used.[4] As with the previous ohmmeter circuits, a test current is derived from a zener diode whose voltage is impressed on a range-switched input resistor of an inverting connected amplifier A_1. This current will flow through the network connected as a feedback element to A_1 but only through R_x if appropriate grounding terminations are made. None of the test current will be shunted away by the other network elements returned to the amplifier input if the voltages on these elements

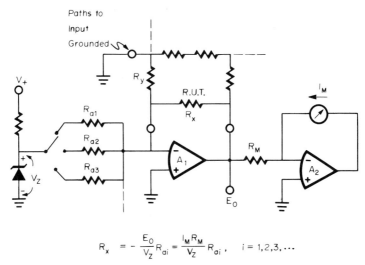

$$R_x = -\frac{E_O}{V_Z}R_{ai} = \frac{I_M R_M}{V_Z}R_{ai}, \quad i = 1,2,3,\ldots$$

Fig. 9.10 Measurement of one resistor in a closed circuit loop is made possible by the virtual ground of an operational amplifier input.

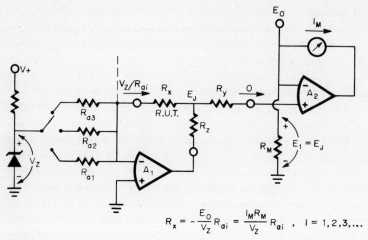

Fig. 9.11 To measure one element of a resistor tee network without access to the junction of the tee, buffering is used to eliminate current flow in one of the tee elements.

are held at zero. This is simply achieved because the virtual ground characteristic of the inverting operational amplifier connection ensures zero volts at the inverting amplifier input. Then, zero voltage can be maintained on network elements that would shunt test currents, such as R_y, by merely grounding the other ends of such elements.

This ensures that all the test current flows in the resistor under test to develop a voltage at the output of A_1 that is directly related to that resistor. Measurement can be made using this voltage E_0 as an output indicator or by converting this voltage to a current for meter drive using amplifier A_2. The resistance of R_x is related by

$$R_x = -\frac{E_0}{V_z} R_{ai} = \frac{I_M R_M}{V_z} R_{ai} \qquad i = 1, 2, 3, \ldots$$

As before, the accuracy of this measurement is limited by the errors of the meter, the zener voltage, the scaling resistors, and R_M. However, in this case the input offset voltage of A_1 can impose greater error. That offset is impressed on input shunting elements such as R_y so that this offset is amplified by an additional gain of $1 + R_x/R_y$. Normally this error will be small unless R_x is very much larger than R_y.

Straightforward measurement of one resistor of a tee network requires access to the junction of the tee, which may not be available in packaged networks or with hybrid thin-film or thick-film networks. With operational amplifier feedback techniques, one element of such a network can be measured independent of the others using the circuit of Fig. 9.11.

Buffering by A_2 is used to reduce the current in one element of the tee network to zero so that the voltage at the junction of the tee can be monitored. This voltage E_J is developed on R_x by the test current; so E_J is directly proportional to the resistance of R_x. To monitor this voltage, buffer A_2 is added. This amplifier can be used as a meter amplifier as shown. Addition of R_M results in a meter current proportional to the value of R_x if no current is then conducted from the E_0 output. The resistance R_x can be found from either signal by the relationships

$$R_x = -\frac{E_0}{V_Z} R_{ai} = \frac{I_M R_M}{V_Z} R_{ai} \quad i = 1, 2, 3, \ldots$$

The accuracy of this ohmmeter response is limited by that of the zener voltage, the meter nonlinearity, resistor tolerance, and amplifier input errors as before. Another limit is imposed by the output voltage range of A_1, which must support the voltage drops on R_x and R_z. For large values of R_z the output of A_1 can saturate with no indication of this condition at the monitored outputs. Where this is possible, an overload indicator[2] should be added to the output of A_1.

9.3 Capacitance Measurement Circuits

Operational amplifier control of test signals also simplifies capacitance measurements with common differentiator configurations. Such configurations are applied to conventional- and null-type measurements in the circuits that follow. With the basic differentiator connection of Fig. 9.12, the virtual ground of the operational amplifier input holds one terminal of a capacitor under test at zero voltage, such that a test signal applied to the differentiator input is impressed totally on the capacitor. Resulting capacitor current is conducted through the amplifier feedback resistor, where it develops a voltage related to the capacitance being tested.

Measurement of the differentiator output voltage defines the capaci-

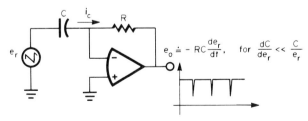

Fig. 9.12 The response of a differentiator to a ramp input voltage is linearly related to the capacitor because the slope of the ramp is essentially constant.

tance value if the effect of the capacitance voltage sensitivity is small. To examine the significance of such sensitivity, the current developed in the capacitor by a test voltage e_r can be expressed by

$$i_c = \frac{dQ}{dt} = \frac{d(Ce_r)}{dt} = C\frac{de_r}{dt} + e_r\frac{dC}{de_r}\frac{de_r}{dt}$$

Note that the first term of this expression is proportional to the capacitance C as desired for measurement but that the second term is not. Fortunately, the second term is negligible for capacitors having other than high voltage sensitivity, and the differentiator output will be

$$e_o \doteq -RC\frac{de_r}{dt} \quad \text{for} \quad \frac{dC}{de_r} \ll \frac{C}{e_r}$$

If the test signal e_r is a ramp, de_r/dt will be constant, and e_o will be a readily measurable dc voltage. Combining a ramp generator with a practical differentiator results in the capacitance measurement circuit of Fig. 9.13.[5] The differentiator formed with A_3 has a gain-limiting resistor R_1 commonly added to preserve frequency stability. Also added is a clamping diode which limits the negative output swing produced by ramp-train reset swing. The clamping makes the average e_o more nearly approximate a dc level proportional to the capacitance.

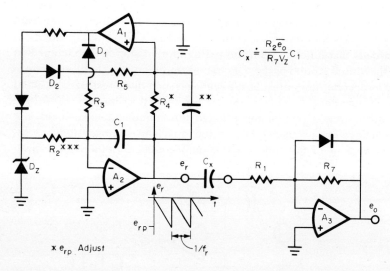

x e_{rp} Adjust

xx e_r Offset Adjust

xxx f_r Adjust

Fig. 9.13 Only three operational amplifiers are needed to form this capacitance measuring circuit.

Forming the ramp-train generator are A_1, A_2, and their feedback elements.[1] Comparator A_1 senses the output of integrator A_2 and supplies a drive voltage to the integrator input in a manner that produces the desired ramp train. When the comparator output is positive, it will bias the zener diode on to apply a constant voltage at integrator input resistor R_2. This voltage results in a negative-going ramp at the integrator output that continues until a comparator trip point is reached. That trip point is established by hysteresis feedback through R_4 and R_5 and defines the ramp-train peak level at

$$e_{rP} = -\frac{R_4}{R_5} V_Z$$

where V_Z is the voltage of the zener diode. At this trip point, the comparator output swings negative, forward-biasing D_1 to drive a small integrator input resistor R_3 for ramp reset. Reset continues until the second comparator trip point is reached. That trip point is at zero voltage since hysteresis feedback is disconnected by the reverse biasing of D_2 in this mode. With the resulting unipolar ramp, polar capacitors can also be measured.

The frequency of the ramp is set by the zener voltage level and the integrator response. By means of D_1 and R_3 the time duration of the positive output integration is made much shorter than that of the negative integration. This produces a ramp instead of a triangle wave and makes the signal frequency essentially determined by the slower integration. In developing the greatest difference in positive and negative integration times, comparator switching time becomes significant and produces a zero level error that must be corrected by selecting a bypass capacitor for R_4. When this is done, the frequency of the ramp train will be

$$f_r = \frac{R_5}{R_4} \cdot \frac{1}{R_2 C_1}$$

From this frequency and the previously defined amplitude e_{rP}, the slope of the ramp is

$$\frac{\Delta e_r}{\Delta t} = \frac{e_{rP}}{1/f_r} = \frac{V_Z}{R_2 C_1}$$

The resulting differentiator output e_o defines the capacitance by

$$C_x = \frac{R_2 \overline{e_o}}{R_7 V_Z} C_1$$

Errors associated with this circuit are primarily due to the dc errors of the differentiator and the response limitations of all three operational amplifiers. During the reset time of the ramp, the differentiator output

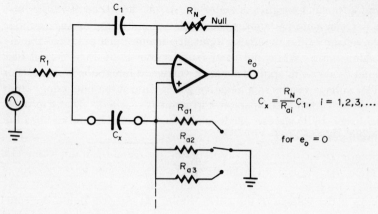

Fig. 9.14 Improved capacitance measurement accuracy with a null technique is achieved with a differential differentiator.

is momentarily interrupted. This reduces the average output voltage measured by a voltmeter readout unless this interruption due to the amplifier response limits is a negligible fraction of the ramp period. A low-frequency ramp reduces this error but simultaneously makes dc errors greater.

Capacitance measurement error associated with test signal deviation and with the nonlinearity of a monitoring meter can be avoided using null measurement techniques. One means of null measurement makes use of the balancing characteristic of the differential differentiator connection as in Fig. 9.14. As shown, a test signal is connected to both differentiator inputs as a common-mode input signal. Such common-mode signals will develop output signals proportional to the difference between inverting and noninverting differentiator gains $\omega R_N C_1$ and $\omega R_{ai} C_x$, respectively. By means of range switching the R_{ai} and variation of R_N, the two differentiator gains can be balanced to null the circuit output. At that null, the two gains are equal and opposite; so the value of C_x is known from the expression

$$C_x = \frac{R_N}{R_{ai}} C_1 \quad \text{for } e_o = 0 \quad i = 1, 2, 3, \ldots$$

While the accuracy of this null measurement technique is independent of the test signal and any meter nonlinearity, it is now limited by potentiometer nonlinearity, reference capacitor accuracy, and amplifier common-mode rejection. Generally, the potentiometer nonlinearity produces the most significant error at about 0.1 percent of full scale. The capacitance of reference capacitor C_1 must be precisely known and must be insensitive to test voltage swing. Signal voltage impressed on the capacitors appears as a common-mode voltage at the operational amplifier inputs, where it produces a common-mode input error related by the amplifier common-mode rejection ratio. That input error signal is amplified by the differentiator gain $\omega R_N C_1$ to produce an output error. To limit this error, the test signal frequency is set low enough to benefit from the higher amplifier common-mode rejection at low frequencies.

9.4 Signal Measurement Circuits

The precise signal processing capabilities of operational amplifiers are applied in the measurement of numerous signal characteristics. Such applications include the voltage, current, frequency, and phase measurement circuits of this section. Other circuits often used in signal measurement include the rms converter, absolute-value circuits, peak detectors, voltage discriminators, and comparators of previous chapters.

9.4.1 Digital voltmeters A digital voltmeter consists of an analog-to-digital converter with gain ranging, a digital readout, and control circuitry. Since low-speed analog-to-digital conversion techniques are suitable for digital voltmeters, these lower cost techniques are commonly used. Among the appropriate analog-to-digital conversion approaches are the dual-slope and voltage-to-frequency converter means considered here. To form digital voltmeters with the voltage-to-frequency converters described in Chapter 8, the converters are connected as in Fig. 9.15 with an output reading counter and control circuits. The resulting digital output lines are then used to drive a digital readout.

More common digital voltmeter circuitry makes use of the dual-slope

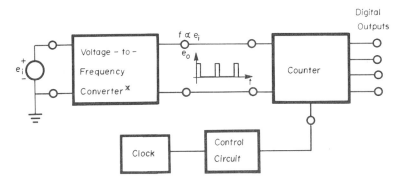

x See Section 8.2

Fig. 9.15 For digital voltmeter applications a voltage-to-frequency converter transforms an input voltage to a related frequency, which can be measured with a digital counter.

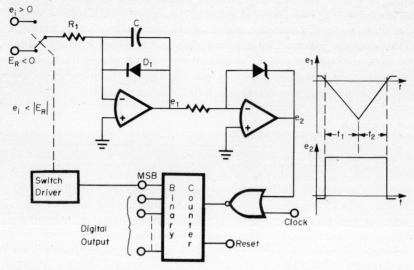

Fig. 9.16 Dual-slope analog-to-digital conversion removes several sources of error from digital voltmeters.

technique, which can be simplified by use of current rather than voltage switching. The dual-slope approach is an improved form of integrating analog-to-digital conversion which compares the time integral of an input signal against that of a reference voltage. In this way several sources of error are removed. As a typical circuit implementation, Fig. 9.16 (Ref. 2) includes an integrator, a zero-crossing detector, a counter that converts integration time to a digital number, and a counter-controlled input switch.

When the counter is reset, the most-significant-bit (MSB) output of the counter causes the switch driver to connect the signal e_i to the input. This causes the integrator output to begin producing a negative-going ramp starting from the level at which it was clamped by D_1. At the zero crossing of this ramp, the comparator switches positive, permitting clock pulses to pass through the NAND gate to the counter. These pulses advance the counter output from its 000 $\cdots$ 00 output state to the 100 $\cdots$ 00 output state, at which point the 1 in the MSB output state causes the input connection to be switched to the reference voltage E_R.

Note that the counter outputs used for the circuit output signal are then all in the zero state again. Thus, the circuit output is essentially reset to begin counting the second time interval t_2. During this interval, the negative reference voltage E_R causes the integrator to produce a positive-going ramp. When that ramp crosses zero, the comparator output swings negative so that the NAND gate blocks clock pulses that would further advance the counter. The counter output remains at that state defining the interval t_2 until reset again.

From t_2 the magnitude of the input signal is defined by considering the associated change in integrator output voltage:

$$\Delta e_i = \frac{-t_2}{R_1 C} E_R = \frac{t_1}{R_1 C} e_i$$

Solving for e_i yields

$$e_i = -\frac{t_2}{t_1} E_R$$

Thus, the signal measured is known in terms of time t_2 defined by the final counter output, the time t_1 defined by the number of clock pulses required to advance the counter from $000 \cdots 00$ to $100 \cdots 00$, and the reference voltage E_R. No measurement dependence upon the integrating feedback elements R_1 and C results, because they determine the integration rates during both t_1 and t_2. These time intervals are dependent upon the clock pulse period, but the measurement relies only on the ratio of t_1 and t_2; so extreme clock pulse stability is not required. In addition, offset in the zero-crossing detector produces compensating effects on t_1 and t_2. As a result of these features, the major sources of error are reduced to the input errors of the integrator, reference voltage errors, and switching time errors.

To avoid error from input switch offset or ON resistance, relays rather than solid-state switches are commonly used in dual-slope digital voltmeters. Alternatively, solid-state switches can be used for improved reliability and economy by employing current rather than voltage switching. Signal and reference currents switched into the integrator input are not affected by switch offset or ON resistance. Current switching is adapted to the above dual-slope digital voltmeter as in Fig. 9.17. As before, an integrator, zero-crossing detector, and counter are used, and these elements perform the same functions previously described. However, the counter MSB output now controls a switched current source to change integrator capacitor current from signal to reference.

The current switched is derived by a controlled current source formed with A_3, Q_3, Q_4, and D_{Z1}.[1] With the current-source connection shown, the current established in R_1 by Q_3 is $(e_i + E_R)/R_1$. A matching current will be developed in the collector of Q_4 for supply to the switches Q_1 and Q_2. Switch Q_2 will be turned off when the counter MSB input is zero during the interval t_1. Then the current supplied to C will be e_i/R_1 as before. At the end of the t_1 interval, the counter MSB input swings positive, turning on Q_2 to add the current from Q_4 to the capacitor C. The

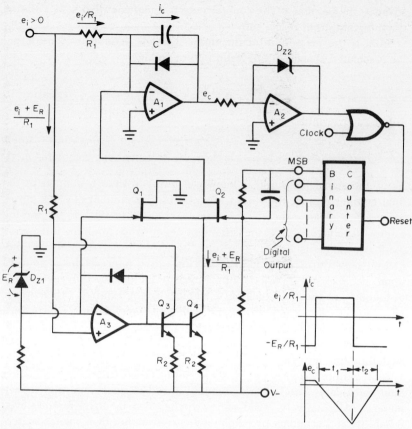

Fig. 9.17 Current switching in a dual-slope circuit provides a higher accuracy digital voltmeter without the need for relays.

net current in C will then be $-E_R/R_1$, just as with the previous circuit. Since the integrator signals are the same as for the circuit of Fig. 9.16, the signal measure is related by the same parameters:

$$e_i = \frac{t_2}{t_1} E_R$$

Response accuracy is limited by the same factors described for Fig. 9.16 plus error introduced by the current source. Again reference voltage inaccuracy and integrator input offsets introduce direct errors. Further error results from the input offset of the current-source amplifier A_3, from any mismatch in the R_1 or R_2 resistor sets, and from mismatch between the current-source transistors. Both emitter-base voltage and beta mismatches between Q_3 and Q_4 will be significant. Accurate transistor matching is required over the full current range defined by $(e_i + E_R)/R_1$.

9.4.2 Ammeters

Elementary ammeters consist of a meter movement with provision for switched addition of shunt resistors that provide range variation. Principal accuracy limitations of such basic meters result from the meter resistance and nonlinearity, and these errors can be avoided with operational amplifier meter realizations. To eliminate the meter resistance error, the meter is simply connected in the feedback loop of an operational amplifier, where it will not develop a voltage in the current path being monitored.

Where the current to be measured is conducted to a power-supply return, a single operational amplifier is sufficient for removing meter resistance error.[1] In these applications amplifier A_1 of Fig. 9.18 is used with terminal B connected to that return point, and the other components shown are not required. Current into terminal A then flows through the meter connected in feedback with A_1, and the meter resistance will result in a voltage at the output of an amplifier. However, the associated voltage between the ammeter terminals A and B is limited to the meter voltage drop divided by the open-loop gain of A_1 plus the input offset voltage of that amplifier. This offset voltage and the input bias current of A_1 are the circuit errors added when using an operational amplifier to remove meter resistance error.

In other applications of ammeters, it is desirable to connect the meter in series with a current path not returned to a common terminal. This requires that the current conducted into one ammeter terminal be returned from the other terminal to avoid interrupting a circuit current path. It is for these requirements that amplifier A_2 and other resistors are added to the circuit in Fig. 9.18. Then, current flowing into terminal A will result in a proportional voltage on the R_{1a} feedback resistor. That voltage is inverted by A_2 to develop a return current from terminal B. Note that

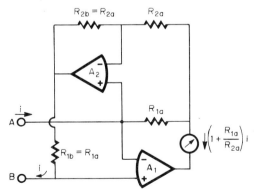

Fig. 9.18 A floating input ammeter is formed by adding an inverter to a basic ammeter circuit.

the inverter formed with A_2 is referenced from terminal A so that voltages present at the meter terminals will not affect the return current. If R_{1b} equals R_{1a}, a current equal to the terminal A input current will be returned to terminal B.

Again, meter resistance will produce negligible error in its feedback connection, but the amplifiers and resistors can introduce measurement errors. The input bias currents of both amplifiers and the input offset voltage of A_2 produce error in the current conducted by the meter. Potentially more serious deviation in meter current can result from the tolerance errors of R_{1a} and R_{2a}, as seen from the meter current expression of the figure. Mismatches between the R_1 resistors and between the R_2 resistors make the current returned from terminal B deviate from that supplied to terminal A. A similar error results because of the input offset voltage and input bias current of A_1. Except for very low current applications, the combined errors of these sources can be made small compared with the nonlinearity error of the meter used.

Where higher accuracy is required, meter nonlinearity error can be avoided using digital voltmeter techniques for ammeters. In the simplest case, a current-monitoring circuit can be used to produce a signal-related voltage input to a digital voltmeter like those of the previous section. Such a voltage output ammeter can be formed with floating inputs suitable for insertion in an ungrounded circuit path as in Fig. 9.19.

This circuit consists of a differential input amplifier formed with A_1 and A_2 and using current feedback,[6] plus an output difference amplifier. Current conducted into terminal A will cause the output of A_1 to swing positive. This in turn produces a current in R_{2a} that is supplied to the

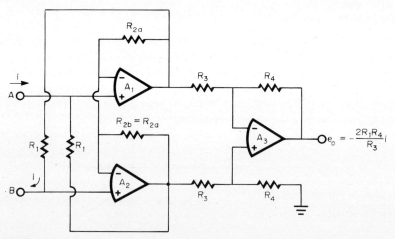

Fig. 9.19 To avoid meter nonlinearity error, an ammeter can be formed with this floating input current-to-voltage converter and a digital voltmeter.

inverting input of A_2, where it drives the output of A_2 negative. That negative voltage conducts the terminal A input current through an R_1 input resistor. Similarly for the opposite polarity, currents conducted out of terminal A result in a negative voltage at the output of A_1 and a positive voltage at the output of A_2. These opposite polarity voltages supply equal and opposite feedback currents to the input terminals through the R_1 resistors. At equilibrium these feedback currents supplied to the circuit inputs equal the signal current.

To supply these input currents, the voltage between the outputs of A_1 and A_2 must be $2iR_1$. Also present at these amplifier outputs is a common-mode voltage induced by the presence of any voltage at the points where A and B are connected. To remove this common-mode voltage, the difference amplifier formed with A_3 is added. The result is a ground-referenced output voltage proportional to the current monitored as expressed by

$$e_o = -2 \frac{R_1 R_4}{R_3} i$$

Measurement of this voltage by a digital voltmeter provides an ammeter indication without the error limitation of analog meter nonlinearity. Measurement accuracy is then determined by the less serious resistor tolerance and mismatches and by the normal operational amplifier errors.

9.4.3 Frequency measurement circuits

An obvious approach to frequency measurements makes use of a digital counter, a storage register, and control circuitry to detect the number of signal repetitions in a given time period. The resulting digital word output can be decoded for display or converted to analog form with a digital-to-analog converter. However, where an analog voltage output is desired, simpler measurement circuits often result with the analog frequency-to-voltage converters described in this section. Described are two such circuits that employ averaging suitable for higher frequency signals or for longer measurement times. A third circuit performs frequency measurement in one cycle of a signal for more rapid frequency measurement.

In the simplest case, signal frequency can be converted to a proportional dc voltage with one operational amplifier. If the signal amplitude is well controlled and if the circuit output is not significantly loaded, such frequency-to-voltage conversion is achieved with the circuit of Fig. 9.20. This circuit differentiates the signal, rectifies the result, and then averages the rectified signal. These operations result in an output voltage of

$$E_o = \left| \overline{-RC_1 \frac{de_i}{dt}} \right| = RC_1 \frac{\Delta e_i}{\Delta t}$$

248 Designing with Operational Amplifiers

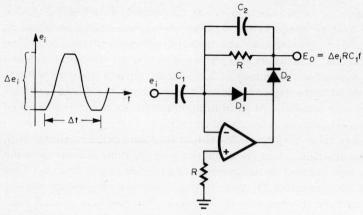

Fig. 9.20 Frequency measurement of controlled amplitude signals can be performed with this simple frequency-to-voltage converter.

where Δt represents the period of the signal or the inverse of the frequency; so

$$E_o = RC_1 \Delta e_i f$$

Thus, the signal frequency is related to the circuit output voltage by

$$f = \frac{E_o}{\Delta e_i RC_1}$$

Note that this relationship is dependent upon signal amplitude Δe_i but not upon signal rate of change. This makes measurement independent of signal waveform, and simple clipping can be used to provide the required stability to the amplitude Δe_i.

Circuit performance relies upon differentiation, rectification, and averaging provided by the various feedback elements. Differentiator feedback elements include C_1 and R, which are chosen to set the circuit gain level. Rectification is performed by the feedback switching of the diodes shown in a manner similar to the basic precision rectifier. That switching is speed-limited by the rate of change of the amplifier output voltage. As a result, measurement bandwidth is limited in the same way as described for precision rectifiers in Sec. 5.4.2.

The averaging function of the frequency-to-voltage converter is provided by a filter capacitor C_2. Choice of this capacitor is basically governed by the maximum allowable output ripple with lower signal frequencies and by desired circuit response to signal frequency changes. In addition, higher values for C_2 can be necessitated by loading on the circuit output. Load currents are conducted from C_2 in the circuit mode where a reverse-biased D_2 disconnects the amplifier output from a circuit load. Such load current drain creates output droop error.

For more general-purpose frequency measurement applications, the frequency-to-voltage converter of Fig. 9.20 above should be preceded by an accurate signal clipper and followed by a buffer amplifier. However, a simpler general-purpose circuit results by combining a one-shot and a low-pass filter as in Fig. 9.21. Each time the input signal crosses the threshold of the one-shot formed with A_1, a pulse of duration t_1 and amplitude V_Z is supplied to the filter formed with A_2. Between such pulses the one-shot supplies a voltage of $-V_Z$ to the filter for some time t_2. The average value of the one-shot output is related to the frequency with which the one-shot is triggered, and that average value is derived by the low-pass filter. Thus, signal frequency can be determined from the filter output voltage by the relationship

$$f = \frac{V_Z - R_5 E_0/R_6}{2 V_Z t_1}$$

Scaling of the expressed frequency-to-voltage converter response is determined by the choice of t_1, V_Z, R_5, and R_6. Circuit operation determining t_1 is described with the circuit of Fig. 6.15 and is expressed by

$$t_1 = (R_1 \| R_2) C_1 \ln \frac{2 V_Z}{V_T} \qquad V_T = \frac{R_2}{R_1 + R_2} V_-$$

As with the previous circuit, the frequency-to-voltage converter of Fig. 9.21 is subject to output ripple that can degrade measurement accuracy

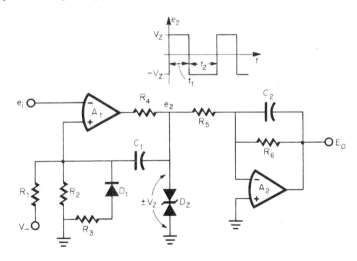

$$E_0 = (1 - 2 t_1 f) V_Z R_6 / R_5 , \quad t_1 = (R_1 \| R_2) C_1 \ln \frac{2 V_Z}{V_T}, \quad V_T = \frac{R_2}{R_1 + R_2} V_-$$

Fig. 9.21 Frequency measurement independent of signal amplitude is provided by a one-shot low-pass filter frequency-to-voltage converter.

250 Designing with Operational Amplifiers

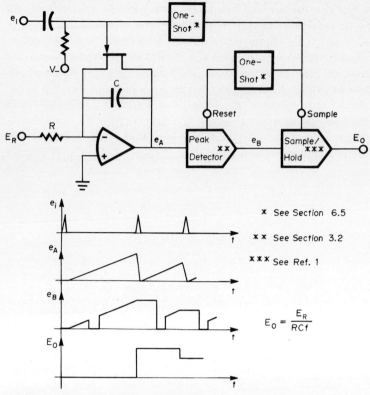

Fig. 9.22 Frequency measurement can be made in one cycle of a signal by detecting the voltage reached by a ramp during the cycle.

at low frequencies. Higher values of filter capacitors reduce this output ripple but similarly increase circuit settling time.

Where more rapid measurements are required, another technique can be applied for frequency-to-voltage conversion in one cycle of the signal. As shown in Fig. 9.22, this technique makes use of an integrator, a peak detector, a sample-hold circuit, and two one-shots. A measure of the signal frequency is derived from the voltage level reached by an integrator before the integrator is reset by an input signal pulse. Because the signal period determines the time between integrator resets, the integrator output produced by a reference voltage in that time will be proportional to the signal period.

To maintain the circuit output voltage at the integrator peak level, the peak detector and sample-hold circuit are added. The integrator peak level is stored in the peak detector, where it is sampled at the end of each pulse by the sample-hold circuit in response to a sample control one-shot.

Following this, the peak detector is reset by a pulse from a second one-shot so that it may acquire the integrator peak level of the next cycle. As a result of this operation, the circuit output remains at a dc level that is inversely proportional to the signal frequency associated with the preceding cycle. Signal frequency is determined from this output voltage by the relationship

$$f = \frac{E_R}{E_0 RC}$$

9.4.4 Phase detectors Measurement of signal phase involves comparison of the signal against a reference signal of the same frequency. Generally, the zero-crossing times of these two signals are compared, and the time difference is used to generate a dc voltage proportional to the phase difference. Phase detectors which operate in this manner are described for square-wave signals and for signals of arbitrary waveform.

When the signal to be measured and the reference signal are square waves, phase detection can be performed with the two-amplifier circuit of Fig. 9.23. This circuit consists of an amplifier A_1 having signal-controlled gain polarity and a low-pass filter formed with A_2. As will be de-

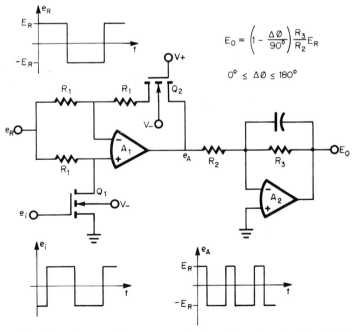

Fig. 9.23 Phase detection of square-wave signals can be performed with switched polarity gain and low-pass filtering.

scribed, the signal e_A passed by A_1 to the filter has an average value determined by the time difference between the two input signal zero crossings. This control of e_A results from the signal-controlled gain polarity of A_1.

The polarity switching of input signal e_i reverses the input amplifier gain polarity by converting it from an inverter to a voltage follower. When e_i is positive, Q_1 is on, grounding the noninverting input of A_1. Then A_1 appears as a unity-gain inverter to the reference signal e_R. The ON resistance of the switch Q_1 produces an input error voltage, but this error is compensated for by connecting a matching ON resistance in the amplifier feedback path. As biased, the compensating FET Q_2 acts simply as a resistor. When Q_1 is off with the negative state of e_i, the reference signal reaches the noninverting input of A_1 direct. No voltage is then across the R_1 summing resistor; so essentially no feedback current flows around A_1. With no feedback current, there is no feedback voltage drop, and the output of A_1 follows the signal e_R. In other words, A_1 has been switched to a voltage follower from an inverter to reverse the polarity of the gain it provides to e_R.

By virtue of this gain polarity switching, A_1 can produce an output signal of nonzero average value from the zero average value signal e_R. If e_i and e_R are in phase, the gain polarity will always be opposite the amplitude polarity of e_R. This results in an output e_A that equals $-E_R$. When e_i and e_R are 180° out of phase, the polarity of the gain provided by A_1 will always be the same as the amplitude polarity of e_R. Then, e_A will equal E_R. For a 90° phase shift, e_A is a square wave of zero average value. Similarly, other degrees of phase shift between 0 and 180° produce other average values at the output of A_1. This average value is detected and inverted by the low-pass filter to result in a phase-related dc output of

$$E_O = \left(1 - \frac{\Delta \Phi}{90°}\right) \frac{R_3}{R_2} E_R \qquad 0° \leq \Delta \Phi \leq 180°$$

Accuracy limitations to this phase detector response result primarily from the amplitude error of e_R, the ripple of the low-pass filter, and the slewing-rate limit of A_1. To reduce output ripple, a large filter time constant is desirable, but the associated longer settling time must also be considered to avoid excessive measurement times. This ripple-settling compromise is typically significant only for lower frequency signals. Higher frequency signals encounter error from the slew-rate limiting of A_1. Because of the finite slewing time of this amplifier, its output signal is not truly rectangular; so its average value deviates from that anticipated. Compensating deviations result if the positive and negative slewing times are equal, but at some point slew-rate limiting will prevent complete output swing during short signal periods.

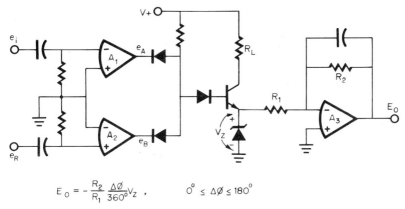

$$E_o = -\frac{R_2}{R_1}\frac{\Delta\phi}{360°}V_Z, \qquad 0° \leq \Delta\phi \leq 180°$$

FIG. 9.24 For signals of arbitrary waveforms, phase detection is provided by signal squaring comparators, an AND gate, and an averaging filter.

Where the input and reference signals are not square waves, the detector of Fig. 9.23 cannot be applied directly. It would be necessary to add signal squaring comparators to the signal and reference inputs. Alternatively, the simpler circuit of Fig. 9.24 can be used with signals of arbitrary waveform. It consists of input squaring comparators A_1 and A_2, an AND gate, and a low-pass filter.

Signal comparison by the AND gate results in the desired output voltage dependence upon signal phase difference. An output voltage is developed that is related to the portions of the signal cycle for which the AND gate output is positive. The gate output is clamped by a zener diode for greater response precision. For a positive gate output voltage, the comparator outputs e_A and e_B must both be positive. If e_A and e_B are in phase, the gate output will be positive for one-half the signal cycle, and a circuit output of $-R_2V_Z/2R_1$ will result. Opposite phase for e_A and e_B holds the AND gate output continuously at zero for a zero output. The phase relationships between e_i and e_R required to produce the above outputs are reverse those required for e_A and e_B. This is a result of the phase inversion provided by A_1, and the net circuit response is

$$E_o = -\frac{R_2}{R_1}\frac{\Delta\Phi}{360°}V_Z \qquad 0° \leq \Delta\Phi \leq 180°$$

Deviations from this response result from factors similar to those described for the previous circuit. Zener voltage variation produces a response gain error; so it is beneficial to add resistor R_L for stabilization of the zener diode current. Output ripple and settling time errors are again determined by the filter time constant. Comparator switching times introduce timing errors in the AND gate output.

9.5 Electronic Thermometers

Electrical measurement of temperature is performed with thermocouples, thermistors, resistance temperature devices, and semiconductor junctions as sensors. For temperatures within the −55 to 125°C range, a semiconductor junction is often the best sensor choice. It is a low-cost sensor with a linear response and error on the order of 1°C over its useful temperature range. For more restrictive temperature ranges, lower sensing error can be achieved. Using bipolar transistors as temperature sensors, thermometer circuits are presented for absolute and differential temperature measurements.

For measurement of absolute temperature, sensing is achieved with a differential transistor pair biased with unbalanced currents[7] as in Fig. 9.25. Unequal currents in the two transistors result in an emitter-base voltage difference of[2]

$$\Delta V_{BE} \doteq \frac{KT}{q} \ln \frac{I_2}{I_1}$$

where K/q equals 0.086 mV/°K and it is assumed that the two transistors are matched. As expressed, ΔV_{BE} is then a linear function of temperature with a sensitivity determined by the ratio of the transistor currents and by physical constants.

To establish the current ratio and to provide a buffered output, the operational amplifier shown is added. Feedback from the amplifier forces the voltage between its inputs to zero so that the voltages on R_1 and R_2 are equal such that

$$\Delta V_{BE} \doteq \frac{KT}{q} \ln \frac{R_1}{R_2}$$

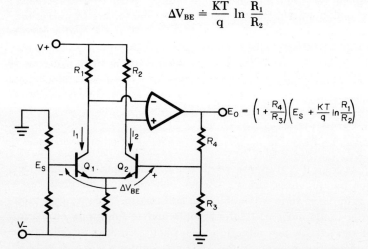

Fig. 9.25 A differential transistor pair with unequal currents provides a linear voltage response for temperature measurement.

This voltage and the scaling voltage E_S are amplified by the gain set with R_3 and R_4 to produce a circuit output of

$$E_O = \left(1 + \frac{R_4}{R_3}\right)\left(E_S + \frac{KT}{q} \ln \frac{R_1}{R_2}\right)$$

As desired, the associated voltage-versus-temperature response is linear. To center the circuit response for zero output corresponding to 0°C or 0°F, the scaling voltage E_S is set.

Measurement accuracy with the thermometer of Fig. 9.25 is primarily determined by the transistor matching and the ratio control of the resistor values. Resistor ratio errors merely produce gain errors, as seen from the last expression. That expression, however, neglects the effects of transistor mismatch, which can also affect thermal response. Close isothermal matching is desired, as provided by matched transistors mounted in a common package. Residual error from the other above sources can be removed through the adjustment of R_1 and R_2 to set the circuit gain.

Response nonlinearity can be introduced by the mismatch in transistor beta drifts, by the input bias current drifts of the operational amplifier, and by use of extreme transistor current ratios. Because of the finite and unequal transistor betas, the transistor emitter currents can have a slightly different ratio than that of the collector currents. This introduces a gain error at any given temperature, but that gain error varies with the nonlinear temperature dependence of beta, producing a response nonlinearity. A similar effect results from the current error introduced by the input bias currents of the operational amplifier. Because these input currents and their thermal drifts also disturb the transistor current ratio, they should be made small to avoid nonlinearity. Similar error could also be produced by the amplifier input offset voltage drift, but use of the transistors as sensors reduces this effect by the transistor voltage gains.

Large transistor current ratios can induce nonlinearity through the associated disturbance of transistor matching. Transistors matched for equal current levels will not typically retain matched thermal characteristics if current ratios over 3:1 are used. With greater current ratios, one or the other transistor will likely be more influenced by high- or low-current density limitations. Since these current density limitations are temperature-sensitive, the errors they produce can vary, producing nonlinearity. With consideration of the various error sources, overall circuit error can be reduced to less than 1 percent.

A variation of the above circuit makes possible measurement of temperature differentials between points. In this case, equal transistor currents are used, and a zero-point scaling bias is not needed, so the circuit becomes that of Fig. 9.26. Differential temperature measurements are made by using the two transistors to sense the temperatures T_1 and T_2 at different

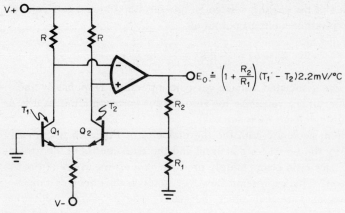

Fig. 9.26 Differential temperature measurement can be performed with differentially connected transistors as sensors.

points. Considering one point as a reference, the relative temperature of a second point is sensed through the approximately linear thermal variation[2]

$$\frac{dV_{BE}}{dT} = \frac{V_{BE} - E_{go}/q}{T} - \frac{3K}{q}$$

$$\doteq -2.2 \text{ mV/°C} \quad \text{for silicon transistors}$$

Then

$$V_{BE2} - V_{BE1} \doteq -(T_2 - T_1)\ 2.2 \text{ mV/°C}$$

That voltage is amplified as controlled by R_1 and R_2 to produce an output voltage proportional to the temperature differential as expressed by

$$E_O \doteq \left(1 + \frac{R_2}{R_1}\right)(T_1 - T_2)\ 2.2 \text{ mV/°C}$$

For this differential thermometer, accuracy limitations are the same as described for the last circuit, with three exceptions: First, the transistor currents are equal; so the matching error previously produced by ratioed currents is avoided. However, the two transistors must be separately packaged to monitor different points; so initial matching may not be so accurate. Additional response nonlinearity results from the slight nonlinearity of the emitter-base voltage thermal variation. As expressed above, this thermal variation is not exactly the constant by which it is approximated.

REFERENCES

1. J. Graeme, *Applications of Operational Amplifiers: Third-Generation Techniques*, McGraw-Hill Book Company, New York, 1973.

2. G. Tobey, J. Graeme, and L. Huelsman, *Operational Amplifiers: Design and Applications*, McGraw-Hill Book Company, New York, 1971.
3. J. Graeme, Single Op Amp Test Circuit Makes Five DC Checks at Once, *Electronics*, August 16, 1973.
4. R. Stitt, Accurately Trimming Closed Resistor Loops, *Electronics*, February 21, 1974.
5. J. Graeme, Getting to Know Capacitors Better with a Simple C/V Display, *EDN*, March 20, 1973.
6. J. Smith, *Modern Operational Circuit Design*, John Wiley & Sons, Inc., New York, 1971.
7. M. Louw, *The Design of Linear Electrothermal Integrated Circuits*, University of Arizona Press, Tucson, 1974.

GLOSSARY

Absolute-Value Circuit A circuit that produces a unipolar output signal equal to the magnitude or absolute value of a bipolar input signal.
Acquisition Time The transition time required by a sample-hold circuit to switch from the HOLD mode to the SAMPLE mode.
Aperture Time The transition time required by a sample-hold circuit to switch from the SAMPLE mode to the HOLD mode.
Bandwidth See "Unity-Gain Bandwidth" and "Full-Power Response."
Bias Current See "Input Bias Current."
Bode Diagram A straight-line approximation to a gain magnitude or phase response curve.
Breakpoint A point at which a Bode diagram changes slope because of a response pole or zero at that point in frequency.
Broadbanding A phase compensation technique in which the compensation applied to an amplifier is reduced for broader bandwidth at higher closed-loop gain levels.
Charge Amplifier An amplifier that produces an output voltage proportional to the charge supplied to its input.
Chopper-stabilized Amplifier An amplifier stabilized against dc drift by chopping the input signal to form an ac waveform that can be coupled to an amplifier through dc isolating capacitors and later demodulated.
Clamping Amplifier An amplifier whose output voltage swing is limited at a specific positive and/or negative voltage level.
Common-Mode Input Capacitance, C_{lcm} The effective capacitance between either input of a differential amplifier and common ground.
Common-Mode Input Resistance, R_{lcm} The effective resistance between either input of a differential amplifier and common ground.
Common-Mode Rejection Ratio (CMRR) The ratio of the differential voltage gain of an amplifier to its common-mode voltage gain.

Glossary

Common-Mode Voltage The average of the two voltages applied to differential amplifier inputs.

Comparator A differential input amplifier used to compare the voltage levels at its two inputs and having high gain so that only small voltage differences are needed to switch the output voltage from one polarity to the other.

Controlled Current Source A current source whose output current is determined by a control signal.

Crossover Distortion Signal distortion occurring at the zero amplitude crossing of a signal.

Current Limiting A means for limiting the output current supplied by an amplifier for protection purposes.

Difference Amplifier An operational amplifier with a feedback configuration that results in an output voltage proportional to the difference of two input voltages.

Differential Input Amplifier An amplifier having two inputs of opposite gain polarity with respect to the output.

Differential Input Capacitance, C_1 The effective capacitance between the two inputs of a differential amplifier when in an open-loop state.

Differential Input Resistance, R_1 The effective resistance between the two inputs of a differential amplifier when in an open-loop state.

Differential Output Amplifier An amplifier having two outputs of opposite gain polarity with respect to a given input.

Differentiator An operational amplifier with a feedback configuration that results in an output signal proportional to the time derivative of the input signal.

Drift See "Input Bias Current Drift," "Input Offset Current Drift," and "Input Offset Voltage Drift."

Droop Rate The rate of change of voltage stored on a holding capacitor like that of sample-hold circuits and peak detectors.

Feedback The return of a portion of the output signal from a device to the input of the device.

Feedback Factor, β That fraction of an output signal fed back to the input.

Feedforward An amplifier phase compensation technique in which high-frequency signals are fed forward around the low-frequency portion of the amplifier.

Frequency Compensation See "Phase Compensation."

Frequency Multiplier A device that produces an output signal of a frequency which is a multiple of that of the input signal applied.

Frequency Response See "Unity-Gain Bandwidth" and "Full-Power Response."

Full-Power Response, f_p The maximum frequency at which an amplifier can supply its rated output voltage and current without significant distortion.

Function Generator A circuit that produces an output signal voltage related to an input signal by some specific or adjustable function.

Gain See "Open-Loop Gain" and "Loop Gain."

Gain Error The difference between the actual closed-loop gain of an operational amplifier with feedback and that predicted by the ideal gain expression.

Guarding See "Input Guarding."

Hysteresis An input-output transfer response lag common to comparators that results in different paths for the two directions of output transition.

Hysteresis Error The separation between the positive-going and negative-going paths of an input-output transfer response curve.

Input Bias Current, I_B The dc biasing current required at each input of an operational amplifier to provide zero output voltage when the signal and input offset voltage are zero.

Input Bias Current Drift The rate of change of input bias current with temperature or time.

Input Capacitance See "Common-Mode Input Capacitance" and "Differential Input Capacitance."

Input Guarding Use of an input shield that is sometimes driven to follow the voltage level of the input signal to remove leakage- and loss-inducing voltage differences between the input path and surrounding stray conduction paths.

Input Noise Current, i_n The equivalent input noise current at each input of an operational amplifier which would reproduce the output noise if the various sources of amplifier current noise were set to zero and if the input noise voltage were zero.

Input Noise Voltage, e_n The equivalent differential input noise voltage of an operational amplifier which would reproduce the noise at the output if the various sources of amplifier voltage noise were set to zero and if the input noise current were zero.

Input Offset Current, I_{OS} The difference between the two input bias currents of a differential input operational amplifier.

Input Offset Current Drift The rate of change of input offset current with temperature or time.

Input Offset Voltage, V_{OS} The dc input voltage required to provide zero voltage at the output of an operational amplifier when the input bias current is also zero.

Input Offset Voltage Drift The rate of change of input offset voltage with temperature or time.

Input Protection Protection of the input of a device from damage due to application of excess input voltage.

Input Resistance See "Common-Mode Input Resistance" and "Differential Input Resistance."

Instrumentation Amplifier A direct-coupled differential input amplifier with internal feedback committed for voltage gain.

Integrator An operational amplifier with a feedback configuration that results in an output signal proportional to the time integral of the input signal.

Integrator Reset Charging of the integrator capacitor to a specific dc level that establishes the initial condition of the integration.

Inverting Amplifier An operational amplifier with a feedback configuration that results in negative voltage gain and thereby inversion of the signal polarity.

Inverting-only Amplifier An amplifier capable only of negative or inverting gain, such as typical single-input operational amplifiers.

Isolation Amplifier An amplifier having high-impedance high-voltage isolation between input and output common returns.

Logarithmic Amplifier An amplifier which develops an output voltage that is proportional to the logarithm of the input signal.

Logarithmic Multiplier A device which derives an output signal proportional to the product of two or more input signals through the use of logarithm-antilogarithm techniques.

Loop Gain, $A\beta$ The gain around a feedback loop formed by an amplifier and its feedback network.

Maximum Selector A device which selects the maximum signal of a set of input signals and connects the selected signal to the device output.

Median Selector A device which selects the median-level signal of a set of input signals and connects the selected signal to the device output.

Minimum Selector A device which selects the minimum signal of a set of input signals and connects the selected signal to the device output.

Multifunction Circuit A circuit having a transfer response that includes multiplication, division, and power or root functions.

Noise See "Input Current Noise" and "Input Voltage Noise."

Noninverting Amplifier An operational amplifier with feedback connected for positive voltage gain, which thereby does not invert the signal polarity.

Offset Current See "Input Offset Current."

Offset Voltage See "Input Offset Voltage."

Open-Loop Gain, A The ratio of the output signal voltage of an operational amplifier to the associated input signal voltage when the feedback loop is open-circuited.

Operational Amplifier A high-gain dc voltage amplifier having high input impedance and low output impedance and capable of developing bipolar output signals from bipolar input signals.

Output Protection Protection of the output of a device from overloads, as commonly provided by output current limiting for an operational amplifier.

Overload Recovery Time The time required for the output of an operational amplifier to return from saturation to linear operation following the removal of an input overdrive signal.

Peak Detector A device which develops and holds an output voltage equal to the peak level of an applied input signal.

Peak Magnitude Detector A peak detector which develops and holds an output voltage equal to the peak level of the absolute value of an applied input signal.

Peak-to-Peak Detector A peak detector which develops and holds an output voltage equal to the peak-to-peak excursion of an applied input signal.

Phase Compensation Frequency response tailoring for stability of a feedback system through the addition of response poles and zeros that reduce high-frequency phase shift.

Phase Detector A device which develops an output voltage level related to the phase difference between two signals.

Phase Margin, ϕ_m The margin by which phase shift around a feedback loop is less than 360° at the unity loop gain frequency. Sometimes it is defined by the margin by which the phase lag of a feedback amplifier is less than 180°, because the negative feedback accounts for the other 180° of phase shift.

Power Booster A buffer amplifier having high current gain and typically unity voltage gain; it is used to boost the output power from an operational amplifier.

Precision Rectifier See "Absolute-Value Circuit."

RMS Converter A circuit that develops a dc output voltage equal in rms value to an input signal of arbitrary waveform.

Sample-hold Circuit A device whose output follows an input signal and then holds the instantaneous value of the signal that existed when the HOLD command signal was applied.

Settling Time, t_s The time required for output voltage settling to within a specified percentage of its final value following a step input.

Signal Conditioner A device that conditions or modifies a signal so as to alter the relationship of the signal with respect to time, frequency, or other signals.

Signal Processor A device that acts on or converts signals by analyzing, routing, rectifying, sampling, etc.

Single-ended Characterized by a single input or output rather than the two of a differential input or output.

Slewing Rate, S_r The maximum rate of change of output voltage when the rated output is being supplied.

Staircase Generator A device which generates an output signal whose waveform consists of successive monotonic step changes in amplitude.

Summing Junction The junction of the feedback and input resistors of a feedback network at which the signal currents from input resistors are summed.

Thermal Feedback Thermal coupling within a device which produces a feedback effect, such as that of operational amplifier input offset voltage change caused by output stage heating.

Two-Wire Transmitter A device which transmits signal in the form of a controlled power-supply current drain so that only two wires need be routed to the device.

Unity-Gain Bandwidth, f_c The frequency at which the open-loop gain of an operational amplifier crosses unity.

Varactor Amplifier A modulated-carrier dc amplifier using the capacitance modulation of varactor diodes in response to low-frequency signals to transmit the signal for ac rather than dc amplification.

Virtual Ground A characteristic of the summing junction of an inverting-connected operational amplifier which resides virtually at ground potential since the very high open-loop gain of the amplifier requires only small summing junction signals to develop the output signal.

Voltage Follower The short-circuit feedback connection of an operational amplifier which results in an output signal that follows the voltage at the noninverting amplifier input.

Voltage-to-Frequency Converter A signal converter which produces an output signal of a frequency related to the signal voltage applied to the device input.

Window Comparator A comparator that detects levels within a set range or window rather than simply distinguishing between levels above and below a set point.

INDEX

INDEX

Absolute-value circuits, 126–148, 258
   differential, 136–143
   precision, 130–143
   response improvement of, 143–148
   simple, 126–130
Ac performance boosting, 1–4, 15–19, 89–93, 143–148
Acquisition time, 78, 89–93, 258
Active filters, 115–121
   digitally controlled, 118–121
   state-variable, 116–118, 121
Adders, 174–177
AGC (automatic gain control), 149–154
Ammeters, 245–247
Amplifiers, 31–56
   bridge, 206–210
   charge, 258
   clamping, 39–46, 73–74, 258
   difference, 175–177, 259
   instrumentation, 31–35, 260
   inverting, 2, 9–10, 20, 175, 260

Amplifiers (*Cont.*):
   motor control, 35–39
   noninverting, 3, 9–10, 21–23, 260
   process control, 200–223
   transconductance, 47–56
Amplitude classifiers (*see* Voltage discriminators)
Analog-to-digital converters, 212–213, 241–244
Aperature time, 258
Arctangent function generator, 195–196
Automatic gain control (AGC), 149–154

Bandwidth:
   multiplier, 190–191
   unity-gain, 261
   (*See also* Full-power response)

Index

Bandwidth boosting:
  absolute-value circuit, 145–148
  multiplier, 190–191
  operational amplifier, 15–19
  peak detector, 89–93
Bias current (see Input bias current)
Bode diagram, 258
Bootstrap feedback, 4–9, 12–13, 72, 90–91, 180, 184–185
Breakpoint, 258
Bridge amplifiers, 206–210
Broadbanding, 258
Buffering:
  input, 4–10
  output, 79–81

Capacitance measurement circuits, 237–241
  null type of, 240–241
Charge amplifiers, 258
Chopper-stabilized amplifiers, 258
Clamping amplifiers, 39–46, 73–74, 258
Common-mode input capacitance, 258
Common-mode input resistance, 258
Common-mode rejection ratio (CMRR), 4, 258
Common-mode voltage, 259
Comparators, 57–76, 259
  hysteresis, 69–76
  one-shot, 171–172
  window, 64–69, 262
Compensation:
  dc error, 1–7
  phase, 91–92, 261
Computing circuits, 174–199
  adders, 174–175
  differentiators, 182–187
  integrators, 177–182
  multiplier/dividers, 187–193
  subtractors, 175–177
  trigonometric circuits, 193–196
Controlled charge reset, 217–223
Controlled current source, 259
Cosine function generator, 194–195
Crossover distortion, 259
Current boosters, 10–12, 14
Current limiting, 259
Current measurement, 245–247
Current sources, 47–56, 200–212
  controlled, 259

Data transmission circuits, 200–223
  two-wire transmitters, 200–212
  voltage-to-frequency converters, 212–223
Dc performance preserving, 1–7
Difference amplifier, 175–177, 259
Differential amplifiers (see Instrumentation amplifiers)
Differential input:
  absolute-value circuits, 136–143
  current sources, 50–56
  differentiators, 184–185
  integrators, 179–181
Differential input amplifier, 21–22, 31–35, 259
Differential input capacitance, 259
Differential input resistance, 259
Differential output amplifier, 259
Differentiators, 182–187, 237–241, 259
Digital control:
  of active filters, 118–121
  of ramp and pulse generators, 163–164
  of voltage regulators, 113
Digital voltmeters, 241–244
Dividers, 187–193
Drift:
  input bias current, 7, 259
  input offset current, 260
  input offset voltage, 2–4, 260
Droop rate, 259
  reduction of, 85–88
Dual-slope converters, 241–244

Error reduction:
  absolute-value circuit, 143–148
  ac, 15–19
  comparator, 69–76
  dc, 1–7
  differentiator, 185–187
  integrator, 181–182
  peak detector, 5, 84–93

Feedback, 259
  bootstrap (see Bootstrap feedback)
  thermal, 229–231, 261
Feedback factor, 16–19, 259
Feedforward, 3, 259
Foldback current limit, 111–112

Frequency compensation, 91–92, 261
Frequency measurement circuits, 247–251
Frequency modulators (*see* Voltage-to-frequency converters)
Frequency multipliers, 121–125, 259
Frequency response (*see* Full-power response; Unity-gain bandwidth)
Frequency-to-voltage converters, 247–251
Full-power response, 259
  boosting, 15–19
Function generators, 259

Gain:
  loop, 260
  measurement of, 227–231
  open-loop, 260
Gain control, 19–30
  bipolar, 22–23
  electronic, 26–30
  polarity, 29–30
  switched, 23–30
  variable, 19–23
Gain error, 175, 177, 183, 189, 260
Glossary, 258–262
Guarding, input, 259

Hysteresis, 259
  error, 69–76, 259
  switching, 60–62

Input bias current, 259
  measurement of, 227–229
Input bias current compensation, 4–7, 85
Input bias current drift, 7, 259
Input capacitance, 259
Input guarding, 259
Input impedance boosting, 4–10
Input noise current, 260
Input noise voltage, 260
Input offset current, 260
Input offset current drift, 260
Input offset voltage, 260
  measurement of, 227–229
Input offset voltage drift, 2–4, 260

Input offset voltage null, 2–4
Input protection, 260
Input resistance (*see* Common-mode input resistance; Differential input resistance)
Instrumentation amplifiers, 31–35, 260
Integrators, 177–182, 260
  reset, 260
Inverting amplifier, 2, 9–10, 20, 175, 260
Inverting-only amplifier, 260
Isolation amplifier, 210–212, 260

Level-triggered one-shot multivibrators, 171–172
Level-triggered oscillators, 62–63
Limiter, 39–46, 73–74
Logarithmic amplifier, 260
Logarithmic amplifier bandwidth, 190–191
Logarithmic multiplier/dividers, 187–191
Loop gain, 260

Maximum selector, 81–84, 94–96, 260
Measurement circuits, 224–257
  capacitance, 237–241
  current, 245–247
  frequency, 247–251
  phase, 251–253
  resistance, 231–237
  temperature, 254–256
  voltage, 241–244
Median selector, 97–98, 260
Meter circuits, 231–247
Minimum selector, 94–96, 260
Modulated-carrier amplifier, 208–210
Motor controllers, 35–39
Multifunction circuit, 191–193, 260
Multiplier/dividers, 187–193

Noise:
  input, 260
  rejection of, 59–60, 200–212
Noninverting amplifier, 3, 9–10, 21–23, 260
Noninverting differentiator, 183–185
Noninverting integrator, 179–180

Offset current, input, 260
Offset null:
   absolute-value circuit, 143–145
   feedforward type of, 2–4
Offset voltage (see Input offset voltage)
Ohmmeters, 231–237
   conventional, 232–235
   for embedded networks, 235–237
   null, 234–235
   for tee networks, 236–237
One-shot multivibrators, 169–192
   level-triggered, 171–172
Open-loop gain, 260
   boosting, 4
Operational amplifier, 261
   test circuits, 227–231
Optical coupler transmission efficiency, measurement of, 226
Oscillators, 149–173
   level triggered, 62–63
   voltage-controlled, 212–223
   Wien-bridge, 149–154
Output boosters, 10–14
Output protection, 261
Overload recovery time, 261

Peak detectors, 5, 59, 76–93, 261
   basic, 76–81
   error reduction, 84–93
   magnitude, 83–84, 261
   reset, 86–88
Peak-to-peak detector, 82–83, 261
Phase compensation, 91–92, 261
Phase detectors, 251–253, 261
Phase-lead compensation, 74–76
Phase margin, 261
Phase-to-dc converters, 251–253
Power boosters, 10–14, 261
Precision rectifiers (see Absolute-value circuits)
Process control amplifiers, 200–223
Pulse generators, 161–164, 212–223
   (See also Timing circuits)

Quiescent currents, measurement of, 227–229

Ramp and pulse generators, 161–164
   digitally controlled, 163–164
   triggered, 161–163

Rectifiers (see Absolute-value circuits)
Regulators, voltage (see Voltage regulators)
Ripple reduction, 102–105
Rms-to-dc converter, 197–199, 261

Sample-hold circuit, 5, 261
Sawtooth generators, 161–164
Settling time, 261
Signal analyzers, 57–98
   comparators (see Comparators)
   peak detectors (see Peak detectors)
   voltage discriminators, 93–98
Signal conditioners, 99–125, 261
   active filters (see Active filters)
   frequency multipliers, 121–125, 259
   voltage regulators (see Voltage regulators)
Signal generators, 149–173
   ramp and pulse generators (see Ramp and pulse generators)
   square-wave generators (see Square-wave generators)
   staircase generators (see Staircase generators)
   timing circuits, 169–172
   triangle-wave generators (see Triangle-wave generators)
   Wien-bridge oscillators, 149–154
Signal measurement circuits, 241–253
   ammeters, 245–247
   digital voltmeters, 241–244
   frequency measurement circuits, 247–251
   phase detectors, 251–253, 261
Signal processors, 261
Sine-function generator, 193–194
Sine-wave generators, 149–154
   single-supply, 149–151
   variable-amplitude, 152–153
   variable-frequency, 153–154
   Wien-bridge, 149–154
Single-ended, 9–10, 261
Slew-rate boosting (see Speed boosting)
Slewing rate, 261
Speed boosting:
   absolute-value circuit, 145–148
   operational amplifier, 15–19
   peak detector, 89–93

Square-wave generators, 154–161
  swept-frequency, 159–161
  triggered, 170–171
  trilevel, 158–159
  zero-based, 157–158
Staircase generators, 164–168, 261
  complementary output, 166–167
  long interval, 167–170
State-variable filters, 116–118
Subtractors, 175–177
Summing amplifiers, 174–177
Summing differentiators, 182–183
Summing integrators, 177–180
Summing junction, 261
Switching regulators, 105–110

Tee networks, 181–182, 185–187
  measurement of, 236–237
Test circuits, 224–257
  amplifier, 227–231
  transistor, 225–227
Thermal feedback, 229–231, 261
Thermal resistance, measurement of, 226–227
Thermometers, 254–256
  absolute, 254–255
  differential, 255–256
Threshold, comparator, adjustment of, 58–60, 72, 74–76
Threshold voltage, measurement of, 225
Time constants, extension of, 118–121, 181–182, 185–187
Timing circuits, 169–172
  level-triggered, 171–172
Transistor test circuits, 225–227

Transmitters:
  two-wire, 200–212, 261
  voltage-to-frequency, 212–223
Triangle-wave generators, 154–161
  swept-frequency, 159–161
  triggered, 170–171
  voltage-controlled, 212–223
Trigonometric functions, 193–196
Two-wire transmitters, 200–212, 261

Unity-gain bandwidth, 261

Varactor amplifier, 261
Vector magnitude generator, 196–197
Virtual ground, 261
Voltage boosters, 12–14
Voltage-controlled oscillators, 212–223
Voltage discriminators:
  comparator, 57–76
  multilevel, 64–69, 93–98
Voltage follower, 8–9, 262
Voltage regulators, 99–115
  digitally controlled, 113
  dual, 113–115
  floating, 202–212
  foldback limited, 111–112
  switching, 105–110
Voltage-to-frequency converters, 212–223, 241, 262
Voltmeters, 241–244

Waveform generators, 149–173
Wien-bridge oscillators, 149–154
Window comparators, 64–69, 262